S16094
621.36   Bone, Jan
BON
     Opportunities in laser
     technology careers
$12.95

DATE			

VGM Opportunities Series

OPPORTUNITIES IN
LASER TECHNOLOGY
CAREERS

Jan Bone

Foreword by
Ted Maiman

VGM Career Horizons
a division of *NTC Publishing Group*
Lincolnwood, Illinois USA

36660000016641

Cover Photo Credits:
Front cover: upper left, Ford Motor Company
photo; upper right, Sharon P. Gregg photo;
lower left, photo reproduced with permission
of AT&T Archives; lower right, Oak Ridge
National Laboratory photo.

Back cover: upper left and lower right, GTE
Laboratories photos; upper right, Richard
Smith photo; lower left, Raycon Corporation
photo.

Library of Congress Cataloging-in-Publication Data

Bone, Jan.
    Opportunities in laser technology careers.

    (VGM opportunities series)
    Bibliography: p.
    1. Laser industry—Vocational guidance.   I. Title.
II. Series.
TA1677.B66   1988   621.36'6023   88-60905
ISBN 0-8442-6514-4
ISBN 0-8442-6515-2 (pbk.)   S16094
                               621.36
                               BON
                                 90/91
                               12.95

Published by VGM Career Horizons, a division of NTC Publishing Group.
© 1989 by NTC Publishing Group, 4255 West Touhy Avenue,
Lincolnwood (Chicago), Illinois 60646-1975 U.S.A.
Library of Congress Catalog Card Number: 88-60905
Manufactured in the United States of America.

  8 9 0 BC 9 8 7 6 5 4 3 2 1

ABOUT THE AUTHOR

Jan Bone has been writing professionally for over 40 years, ever since, as a 16-year-old, she had her first newspaper job on the Williamsport, Pennsylvania, *Sun*. Her bachelor's degree is from Cornell University, and she holds an M.B.A. degree from Roosevelt University.

When Ted Maiman successfully invented the first working laser in his laboratory at Hughes Aircraft Corporation, Jan was busy caring for four sons—the oldest in kindergarten. During the three decades that followed, however, her interest in technology has grown significantly.

From 1977 to 1985 Jan served as an elected member of the Board of Trustees of William Rainey Harper College in Palatine, Illinois, and from 1979 to 1985 she was board secretary. During that time, the college set up a CAD/CAM Center to train students and professionals in computer-aided design and computer-aided manufacturing, sparking her interest in the field.

She has studied production management and is an associate member of the Society of Manufacturing Engineers and two of its divisions: Computer and Automated Systems Association (CASA) and Robotics International (RI). Because of her interest in lasers and laser technology, she is also a member of SME's Laser Council.

A prolific free-lance writer, Jan is senior writer for the *Chicago Tribune*'s special advertising sections and has written special features sections material for the *Chicago Sun-Times* since 1982. She has written for publications as diverse as *Bank Marketing, Family Circle, Woman's World, Medical Tribune,* and *National ENQUIRER.* In 1988 she was named contributing editor of *Bank Administration,* the magazine of bank management.

She is coauthor with Ron Johnson of *Understanding the Film* (National Textbook Company, Lincolnwood, Illinois, 1985) and author of *Opportunities in Film Careers, Opportunities in Cable Television, Opportunities in Telecommunications, Opportunities in Computer-Aided Design and Computer-Aided Manufacturing,* and *Opportunities in Robotics Careers* in the VGM Career Horizons series.

Jan has won the Chicago Working Newsman's Scholarship, the Illinois Education School Bell Award for Best Comprehensive Coverage of Education by dailies under 250,000 circulation, and an American Political Science Award for Distinguished Reporting of Public Affairs. Since 1983, she has been listed in *Who's Who of American Women.*

She is married, the mother of four married sons, and grandmother of Emily Diane.

ACKNOWLEDGMENTS

The following individuals were especially helpful in the development of this book: Gary Benedict, Edesly Canto, Patrick Doolan, Ellet Drake, Jack Dyer, Robert Ford, Gordon Gould, Joe Hlubucek, Susan Hicks, Reena Jabamoni, Rick Jackson, Frank Jacoby, Tung H. Jeong, James Johnson, Alan J. Jones, Hamid Madjid, Fortuneé Massuda, Vivian Merchant, George L. Paul, Jeri Peterson, Judith Pfister, Robert Prycz, Greg Rixon, Howard Rudzinsky, John Ruselowski, George Sanborn, Fred Seaman, M. J. Soileau, Doris Vila, and Carol Worth.

Special thanks to Theodore H. Maiman.

The author also acknowledges the assistance of the British Information Service, ABET, Alberta Laser Centre, American Association of Engineering Societies, American Society for Laser Medicine and Surgery, Australian Embassy, Battelle Laboratories, Bell Labs, Edison Foundation, Electronic Engineering Associates, Gas Research Institute, Hewlett-Packard, Institute of Industrial Engineers, Lake Forest College, Laser Focus/Electro-Optics, Laser Institute of America, Lawrence Livermore National Laboratories, Optical Society of America, Raycon Corporation, School of the Art Institute of Chicago, Society of Manufacturing Engineers, Society of Women Engineers, SPIE, University of Arizona Optical Sciences Center, University of Central Florida/CREOL, University of Rochester Institute of Optics, and Westinghouse.

FOREWORD

It takes courage to work on something new and different—not to be part of the majority of people who are comfortable doing things in a familiar pattern. When I made the first laser in 1960, I found out first-hand how satisfying it can be to stay with an idea you believe in.

From the public relations standpoint, the press announcement of that laser was extraordinarily successful. The news hit the front page of every major newspaper in the United States and of many papers overseas. Unfortunately, a typical headline read, "Los Angeles Man Discovers Science Fiction Death Ray!"

Shortly thereafter, the owner of Knotts Berry Farm, a popular amusement park, phoned. He wanted to use the laser in a shoot-the-duck game. A representative of the Ice Capades wanted laser light for the spotlights on his performers because of its purity. The president of the American Meat Packers Association wanted to use a laser to stun hogs.

Here were three forward-thinking, progressive, entrepreneurial people who looked on lasers as a tool they could use creatively in their work. Whether it was feasible or not was beside the point. They were ready to try.

Nearly 30 years later, the laser is still as exciting and marvelous as it was in its "gee whiz!" days. I'm no longer pulled aside at scientific conferences and asked, "Do you think the laser is ever really going to be useful?" Instead, when someone reads my name badge, they may say, "My grandmother's eyesight was saved because of laser surgery," or even, "Thank you for my job."

The acceptance of lasers came rather slowly, just like that of the airplane and the auto. I seriously doubt that the Wright brothers ever dreamed there would be flights around the world and planes carrying hundreds of passengers. And for many years, automobiles were looked on as a rich man's toy.

I certainly didn't envision lasers becoming part of everyday life ... as familiar as the supermarket checkout scanner or the compact audiodisc.

From the beginning, I thought lasers could be used in medicine, but I didn't dream how fantastic today's results would be or the difference they would make in diagnosis and surgery. I felt lasers would be used in communication, but I didn't really see how. It wasn't practical until low-loss fiber was invented ten years later.

Today, lasers in manufacturing improve yield and productivity. You can cut through half an inch of steel with a laser faster than just about any other way.

Today, lasers are used for computer printers and extremely high-density information storage. And tomorrow, tiny lasers inside a computer will achieve much faster speed and performance with optical computing than digital computers now offer.

It's even conceivable that lasers could be the key to solving the energy problem through laser fusion!

To me, lasers are one of the fastest growth industries in the world. There will be jobs in laser technology, not only in working with current laser applications, but also jobs that don't yet exist. The courage of young people—people like yourself who are not afraid to go ahead with ideas they believe in, despite possible discouragement—will make those jobs possible.

Books such as this, which help introduce young people to lasers, play an important role in stimulating imagination and creativity. I hope the laser brings as much satisfaction to your life as it has to mine.

Ted Maiman
Inventor of the first laser

CONTENTS

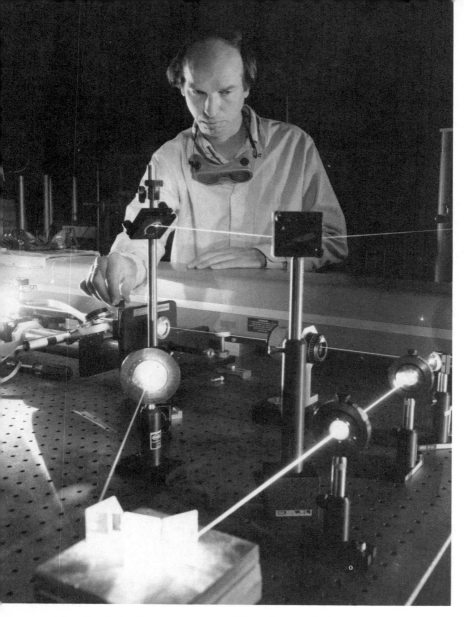

A laser is a device that generates or amplifies coherent radiation at frequencies in the infrared, visible, or ultraviolet regions of the electromagnetic spectrum. (Reproduced with permission of AT&T Archives)

CHAPTER 1

LASERS: AN IMPORTANT TECHNOLOGY

When laser expert M. J. Soileau speaks to fourth-grade students, he takes a laser along. The small black box, about 8 inches long and 1 1/2 inches wide, fascinates the youngsters, who crowd around him. But when he tells them it's a laser, they don't believe him.

"How come you can't see the beam?" they ask, puzzled. To them, lasers mean the lightsticks used by Luke Skywalker and Darth Vader for their duel in *Star Wars.*

Soileau, who heads University of Central Florida's CREOL, the Center for Research in Electro-Optics and Lasers, patiently explains to the students why light doesn't show up in an unpolluted atmosphere. If they want to see the beam, he tells them, he can clap two blackboard erasers together, making chalk fly through the air. A light beam aimed through the dust cloud will then be noticeable.

The second question he invariably gets from a grade school class is, "Don't lasers make holes in things?" Soileau explains they can—and do, especially in industrial or medical applications. Many lasers, however, he tells them, don't have enough power to do anything except to appear bright.

Despite the misconceptions about lasers that many of us hold, these devices fascinate and intrigue us. The very notion of lasers suggests images of speed, brilliance, and modern technology. Laser surgery sounds appealing and desirable to many prospective patients. Laser printers, teamed with computers and software, are helping many small (and large) businesses with desktop publishing. And, in an attempt to cash in on the growing popularity of lasers and the image of quickness they convey, one Chicago suburban firm offering limousine service to O'Hare Airport even calls itself "Laser Livery."

Whether adolescents are playing Lazer Tag or having phone conversations through fiber-optic cables that use lasers, lasers are becoming more familiar to young people. In fact, you may "see" a laser in use several times a week, though you may not realize it. The supermarket checker who runs your purchases over a scanning device, which flashes a signal to the cash register and the computer that runs it, is using a low-intensity helium-neon laser to "read" the Universal Product Code striping.

From tiny lasers like the one at the cash register to the world's biggest laser, which is housed in a building about the size of an NBA basketball arena at California's Lawrence Livermore National Laboratory, lasers are playing an ever-increasing role in our lives. Yet the first working laser was not invented until 1960. In fact, at one time, lasers were called "a solution in search of a problem." Unlike many inventions that were developed to answer perceived needs and solve perplexing, real-world problems, the laser was invented before there were any practical applications for it. Those only came later. Scientists, industry researchers, and technicians still are finding new ways in which lasers can help.

WHAT ARE LASERS?

The word *laser* itself is an acronym—a word formed from the initial letters of the successive parts of a term. *Laser* stands for *l*ight *a*mplification by *s*timulated *e*mission of *r*adiation. A laser is a device that generates or amplifies coherent radiation at frequencies in the infrared, visible, or ultraviolet regions of the electromagnetic spectrum.

That's a complicated definition, especially if you are not familiar with physics or technology. Chapter 2 of this book explains more about how a laser works. It also tells the story of the invention of the laser by Theodore H. Maiman, who in 1960 was a research scientist at Hughes Aircraft. Maiman calculates that the whole nine-month project that resulted in the first operating laser cost Hughes no more than

$50,000, including his salary, his assistant's salary, and overhead—certainly a bargain.

Contrast that small expenditure by Hughes Aircraft with today's economic impact of lasers. *Laser Focus/Electro-Optics,* a leading trade publication, has reported 1987's worldwide sales of commercial lasers at $570 million and projects an 8 percent growth (to $613 million) for 1988.

Basically, a laser is a unique way of amplifying light, making it brighter and brighter. Laser light is different from ordinary light, such as that produced by the sun or by an incandescent light bulb. Lasers are useful for many applications, because laser light has special properties: monochromaticity; a narrow, columnated beam; and coherence.

THE LASER INDUSTRY

Laser Focus/Electro-Optics, a major publication in the laser field, keeps close track of laser sales and use. Results of its research are reported annually—a short report in the January issue of the magazine and an expanded version in *The Annual Economic Review and Outlook.* Information on purchasing the longer report is available from *Laser Report,* P.O. Box 1153, 119 Russell Street, Littleton, MA 01460-0753.

According to the magazine, in 1987 materials processing represented 27 percent of the total market for lasers—approximately $129 million in sales. The medical laser market, which uses carbon dioxide lasers heavily, was up to $81 million, with 1988 sales projected at $91.4 million, a growth rate of 12 percent. *Laser Focus/Electro-Optics* estimates the military laser market to be over $100 million for 1988. Other areas of laser applications projected to grow significantly in sales volume for 1988 were diagnostic medicine, research and development, bar code scanning, and optical memories.

THE PATENT CONTROVERSY

Roughly 60 percent of all commercial laser sales are for gas discharge lasers. And in 1987 an important legal victory affecting those sales was won by Gordon Gould, the inventor who was awarded the patent covering those lasers, after a lengthy struggle.

"I conceived the laser in late 1957," Gould remembers, "and I should have applied for a patent right away, but I thought (wrongly, as it turned out) that I had to have a working model first. I left Columbia University, where I was a graduate student, and joined a company where I thought I could get a laser built. Although I applied for a patent in April 1959, by that time there were other inventors applying for patents on various aspects of lasers."

As Gould explains, the U.S. Patent Office has a procedure for deciding who has the right to a patent where claims overlap in applications—a situation the Patent Office calls "interference."

"My application contained many inventions, including two different types of lasers, and it covered various other aspects of lasers. Consequently, there were five interferences with other inventors."

One of those inventors was Charles Townes, a professor at Columbia University, where Gould was a graduate student. In 1951 Townes, continuing experiments first begun in Germany by other researchers, suggested separating a beam of ammonia molecules into two portions. The molecules in each portion of the beam did not have the same energy states; in one portion, the energy state would be higher. Early scientists studying quantum mechanics, a particular branch of physics, had believed that if an electromagnetic beam with a particular resonant frequency were passed through a medium, molecules of the beam in a higher state of energy might be stimulated to fall to a lower state of energy—and in the process might reinforce the primary beam.

Townes used a microwave oscillator in his experiments and passed the high-energy portion of his ammonia beam through a cavity that resonated at the frequency that matched the energy difference between the high- and low-energy states. Eventually Townes was awarded a patent for his *maser,* a word coined as an acronym for *m*icrowave *a*mplification by *s*timulated *e*mission of *r*adiation. Masers and lasers are theoretically similar, but masers operate at frequencies in the microwave region of the spectrum, while lasers operate in the light range of the spectrum. Later on, the Townes patent was licensed to laser manufacturers.

Meanwhile, at Bell Laboratories Towne's brother-in-law, Arthur Schawlow, was continuing research on optical masers. Together, Schawlow and Townes proposed a way to get optical maser action. Their plan called for an alkaline vapor to be placed in an optical cavity to serve as an active medium. Such a medium, they felt, could be excited in such a way that if an optical wave were present, it would be amplified as it moved through the medium. According to Gould, Schawlow and Townes didn't realize the active medium could be excited by light.

The work of Townes and Schawlow eventually led to the awarding of another patent in 1960, U.S. Patent Number 2,929,922. Because Schawlow and Townes had applied for a patent before Gould's original application, they were considered the "senior parties."

"I was unable to prove the diligence required to establish a date for my work that was earlier than their patent application," Gould says. "They thought they had won." What the Schawlow and Townes patent claimed to cover, Gould says, was the resonator—the pair of mirrors required to shine the light back and forth through the laser amplifier.

Gould, however, had by no means given up his fight to be recognized for his work with lasers and to have his claims recognized by the U.S. Patent Office. It was a long and difficult struggle. In fact, the Patent Office eventually required Gould to divide his original application into six different applications.

"What took up the time," Gould says, "was dealing with these interferences. Each took several years to resolve. They were run in sequence, rather than at the same time, so the determination on the last one was not completed until 1973, 14 years after I'd originally filed. By that time, the industry had virtually forgotten there was such a thing as the Gould patent applications."

Complicating the matter further was the fact that the laser industry had become large and mature by 1977, the year in which Gould was issued his first patent on amplifiers. By then there were many companies, many laser products, and many ways in which lasers were being used. "The amount of money involved for royalties was big," Gould says, "big enough so that nobody was going to just write me a check if I called them up and told them about my patents."

What Gould's two patents covered, he says, were two different kinds of amplifiers that built up the strength of the light beam. One amplifier was the so-called optically pumped laser, consisting of a rod of appropriate material, such as ruby, with a flashlamp beside it. The light from the flashlamp excites the ruby to a state where it serves as an amplifier.

Nearly one-third of all lasers today, Gould says, are optically pumped—making that patent a significant one.

In 1979 Gould was issued a second patent. That patent covered the use of lasers for several different kinds of processes that require heat: welding, heat treating, evaporating materials, and similar chemical reactions.

The third patent issued to Gould, in November 1987, is probably the most important of all. It covers a different kind of amplifier, the discharge laser. Close to 60 percent of all commercial laser sales are of that type, according to Gould. When this patent was issued, U.S. Senator Arlen Spector of Pennsylvania held an award ceremony and press conference honoring Gould. Since Senator Spector heads the subcommittee that oversees the U.S. Patent Office, his support and acknowledgment of Gould were significant.

"The awarding of patents was established in the U.S. Constitution," Gould says. "Our forefathers felt there should be a patent system to give an inventor some rights to his invention ... to encourage inventors to get the inventions out, instead of holding them secret. Yet it's clear that if it takes 28 years to get a patent issued (from Gould's original application in 1959 to the awarding of the discharge laser patent in 1987), the system is not working as intended."

Gould received another laser patent in 1988—a patent he refers to as "the Brewster angle window patent." Some laser tubes have slanted windows on the end. When those windows are at a particular angle (Brewster's angle), laser gas can pass through them without any loss of power. "That's an important development in discharge lasers," Gould explains, "because the gases have to be contained in the laser tube, and you want to get the beam through the tube without any power loss."

Although the CO_2 laser (with projected total commercial worldwide 1988 sales of $106.9 million, according to *Laser Focus/Electro-Optics*) is covered under his 1979 patent for the discharge laser, Gould says that "the more patents you have, the less likely it is that someone will attempt to overturn them."

WHY THE GOULD PATENTS ARE IMPORTANT

Since the beginning of the laser industry, the question of who had the rights to license lasers (because they owned the patents) has been controversial. Gould, of course, has contended for nearly 30 years that he should be recognized and awarded patents for his work.

Townes was awarded his patent for masers and won a Nobel Prize in 1964 for his achievements with the ammonia maser and subsequent developments in masers and lasers. Schawlow and Townes received a laser patent in March 1960.

Shortly thereafter, Gould precipitated an interference proceeding in the U.S. Patent Office. Although the Patent Office later decided the Schawlow-Townes patent had been properly awarded, the patent expired (as all patents do) 17 years after it was issued. Gould received his first patent (U.S. Patent Number 4, 053,845) for "Optically Pumped Laser Amplifiers" on October 11, 1977.

With patents in hand and with a strong determination to win what he felt was due him, Gould sued various laser companies. It was not an easy legal battle; one of the patent infringement suits dragged on for more than 10 years. A significant victory in 1987 against one of the leading laser companies was pivotal, he says, since it showed laser companies that even a properly defended suit (Gould's description) was lost. "That verdict showed other companies in the laser industry that my patents were valid," Gould says.

At stake eventually: penalties from laser companies found to be infringing on the Gould patent and royalties from more than 200 laser manufacturers—royalties that Gould expects to rise to $20 million per year. As a result of winning the "pivotal" suit, Patlex Corporation, which owns a 64 percent interest in the Gould patent, has been signing up licensees. The list includes, Gould says, manufacturers, sellers, and users, including IBM, General Motors, AT&T, Ford, and Chrysler.

Clearly opportunities in laser technology—jobs you may hold—are related to the financial health of the companies involved with lasers, a health that may hinge on their obligations with respect to the Gould patents. The years ahead will be interesting as these economic and legal questions are resolved.

CHAPTER 2

HOW LASERS WORK

There are different kinds of lasers. Some are solid-state. Others use gases, such as helium and neon (He-Ne), argon, krypton, or carbon dioxide (CO_2). Ion lasers are used in the printing industry, in therapeutic and diagnostic medicine, and in light shows in entertainment. Diode (or semiconductor) lasers are becoming more popular, especially for optical disks and communications uses. Dollar sales are projected to rise for dye lasers (from $17.4 million in 1987 to $22.4 million in 1988) and for excimer lasers (from $22.4 million in 1987 to $24.9 million in 1988), according to *Laser Focus/Electro-Optics.*

Regardless of the kind of laser, however, all lasers work on similar principles. They are devices that generate or amplify light.

PARTS OF A LASER

Lasers generally have four parts: (1) an *active medium* made up of atoms, molecules, ions, or a semiconducting crystal; (2) an *excitation mechanism* that excites the atoms, molecules, ions, or semiconducting crystal into higher energy levels than their normal state; (3) *elements* that let radiation bounce back and forth over and over again through the active medium, amplifying the light; and (4) an *output coupler,* a special mirror at one end of the laser that is constructed in such a way that some of the laser light escapes from the active medium.

One argument used unsuccessfully in a lawsuit about the Gould patent was that laser light occurs in nature. The laser manufacturer suggested that sunlight stimulating the atmosphere of the planet Mars was causing a lasing action. The Martian surface acted as a highly reflective mirror, and the interface between space and the Martian atmosphere acted as an output mirror. Since the components common to all lasers (an energy source, something being lased, and two mirrors) existed as a natural phenomenon, the manufacturer said, lasers should not be patentable. However, the courts did not buy this argument and have upheld the validity of the Gould patents.

"Just about anything can be stimulated," says Gary Benedict, chairman of the Laser Council of the Society of Manufacturing Engineers. "The Americans have made a laser out of Jell-O, and the Russians, out of vodka."

HOW LASERS AMPLIFY LIGHT

Regardless of just what the active medium is, however, the purpose of the excitation mechanism is to excite the electrons or ions the medium contains. Scientists think of electrons as traveling around the nucleus of an atom in various orbits. When an electron is excited, the electron jumps to an orbit with a higher energy level. When it returns to the ground state, it gives off energy in the form of a tiny bundle of electromagnetic energy called a photon. If the photon comes near another electron from a different atom—an electron that is in this persistent higher energy state—the photon can induce the premature transition of the second electron so that it too gives off a photon. "One photon stimulates the in-step emission of the second photon," explains Dr. Hamid Madjid, associate professor of physics at Pennsylvania State University.

"Each of these two photons can pass another electron and release it, so pretty soon you have 4 photons, and 8 photons, and 16 photons, and so forth, and you start generating coherent photons.

"This occurs either in a glass tube filled with a mixture of gases, or in a solid material, such as a ruby rod. If those coherent photons move in the axis of the tube or the rod, they induce stimulated emission of more photons.

"On one side of the laser there is a reflecting mirror; on the other, a semitransparent mirror that lets a little light through. The reflecting mirror bounces the small amount of laser light back into the active medium, where it is amplified again and again. This mirror is a special mirror that reflects almost all of the laser light that strikes it. The second mirror, called the output coupler, lets much of the light reflect back into the tube but also lets some of the amplified light escape. This amplified light is called the laser beam."

"PLAYERS" IN THE RACE

Lasers are so common today that it seems as if it would not be hard to build one. But that was not the case 30 years ago.

In 1959 Ted Maiman, a 32-year-old research scientist at Hughes Aircraft Company in Malibu, California, decided to test for himself the accuracy of measurements by another scientist in a previously published paper in order to see whether a ruby crystal was really as inefficient as had been reported. The maser already existed. Would it be possible to produce something similar, using optical, rather than microwave, frequencies to stimulate the emission of radiation and amplify light?

"The race was on," Maiman remembers. "Universities and major research labs wanted to be first to make a laser." Among them were Bell Telephone Labs; the Radiation Laboratory at Columbia University; the Princeton, New Jersey, labs of RCA; the Schenectady Research Lab of General Electric; the IBM Labs; and the Lincoln Laboratory of MIT—all significant 'players.' Other researchers were working in Germany, Japan, and Britain."

Although other scientists were working hard to be "first," Maiman had a somewhat unusual background, a background that he feels may have been partly responsible for his success in making the first working laser. "What was needed," he says, "was a combination of disciplines and experience.

"I'd gotten my Ph.D. in physics. I was an experimental physicist. I knew the theory and the concepts in physics behind the idea of lasers.

I also had practical lab experience, as well as a background in electronics—plus intense motivation and drive."

Maiman's dissertation, completed four years previously, included work in microwaves and optics. He knew about masers and had worked with them, but he felt they wouldn't prove to be practical, since the maser required cooling to within a few degrees of absolute zero. Generating *coherent* light by the concept of stimulated emission sounded more feasible to him.

WHAT "COHERENCE" MEANS

Coherence is one of the unique properties of laser light that makes it so valuable and so important in many laser applications. As Madjid explains to students, coherence can be thought of as an ordered phenomenon.

Light from an incandescent bulb (the familiar electric light) is incoherent. The light waves coming from that bulb differ from each other in frequency and wavelength, direction, and phase.

These are difficult concepts to understand if you haven't studied physics. Let's talk about them, one at a time.

Scientists believe that light travels in waves, just like waves on the ocean. Each wave has a high point, called a peak or crest, and a low point, called a trough. If you were standing on a platform in the ocean, holding your hand out horizontally, as each wave passed by, the crest of the wave would touch your hand. The number of times that happened—that is, how many crests went by in each second— is called *frequency*. Frequency is measured in units of reciprocal time— that is, 1/second.

Ordinary light (coming from an incandescent bulb, or even the familiar sunlight) is jumbled up, and its light waves vary in frequency. In contrast, all the waves of laser light are identical in frequency, since they originate from identical atomic transitions. The word scientists use to describe frequency coherence is *monochromaticity*.

Scientists have discovered there is an electromagnetic spectrum of varying kinds of energy. The oscillations and waves within this spectrum produce both electrical and magnetic effects. Most of these energies are invisible. We can measure these waves, characterizing them by frequency and wavelength. For convenience, we divide the electromagnetic spectrum into sections, classified by the ways in which these energies are generated and used. The electromagnetic waves that have the lowest frequencies are radio waves. Next are the microwaves. Above the microwaves (but still invisible to us) are the infrared frequencies. We can feel these as heat, even though we do not see them.

Visible light is an extremely narrow part of the electromagnetic spectrum. The colors we know range from red, starting at a wavelength of 760 nanometers, to violet, which ends at a wavelength of 360 nanometers. (A nanometer equals one-billionth of a meter.) Above visible light on the electromagnetic spectrum come ultraviolet light, X rays, and gamma rays.

The "white" light from an incandescent bulb or the sunlight that we "see" is really made up of many different colors. You can prove this yourself by looking at a beam of light before and after it is passed through a prism. The prism breaks the light up into its different colors.

Laser light contains light of virtually only one color. This color can vary, of course. In fact, lasers can produce frequencies and wavelengths that range from those of infrared rays through visible light and into ultraviolet light. The light from any one laser is concentrated into an extremely narrow band of frequencies.

In the ocean the distance from one wave crest to another is called the wavelength. It is measured in units of distance, such as feet or meters. Ordinary light is made up of waves of differing lengths. Laser light, however, has waves of the same wavelength; they are identical.

Direction and phase are also important concepts. Imagine a group of people who do not know each other walking across a bridge. They are not all moving in step. Some go one way, while others go another. Some people are a little ahead; others lag behind. Ordinary light has wave patterns like this group of people. It is called *spatially incoherent.* The waves do not come at regular intervals. They are not all moving in the same direction.

Laser light is different. Laser light has spatial coherency. Waves of laser light are somewhat like a large, well-disciplined marching band crossing the bridge—each row exactly the same distance from the row in front and in back. Because the band members are in step, every one of those hundreds of people puts his or her foot down at the same time, generating far more force than the random walkers. Laser light works the same way. Because it is spatially coherent, the electric field reaches a maximum for all the little wavelengths at the same time, acting in unison. It is the coherence of laser light that gives it such power.

MAKING THE FIRST LASER

By 1960, when Maiman was a Hughes researcher, scientists were used to working with coherent light from radiation: ordinary radiowaves, the AM radio, the FM radio, VHF (where television is on the electromagnetic spectrum), UHF (the higher bands of television), and even microwaves. If coherent light could actually be generated—and some leading scientists of the time believed this would never be possible—it would be an important step and a breakthrough, since light waves are about 10,000 times higher in frequency than the microwave section of the spectrum.

Maiman wanted to try.

He considered using potassium vapor, an idea Townes and Schawlow had suggested. He considered using an electrical discharge similar to that in a neon sign—an idea which other researchers had proposed. However, what he really wanted, he decided, was a simple, rugged, solid material—one that was fluorescent, so its crystals would glow under ultraviolet light. Since he was already familiar with the optical properties of synthetic ruby crystal from his earlier work with masers, Maiman chose the ruby for his experiments.

One problem with using the ruby, however, was that ruby light might not be efficient. A paper had been published by another scientist, suggesting that if you stimulated the ruby by shining ordinary (incoherent) ultraviolet, green, or blue light on it, the ruby would glow red. Nevertheless, according to the published paper, the efficiency of that fluorescence was extremely low. As Maiman remembers, the paper predicted that only 1 percent of the energy would be released in the red glow; the remaining energy would be absorbed in the ruby by the green light.

Maiman's own research had already showed him that to get atoms in the ruby crystal excited enough to obtain coherent laser light, he would need an extremely intense light to get the process going. An efficiency of only 1 percent would make using the ruby crystal impractical.

Consequently, Maiman abandoned the ruby idea. "I had no reason to doubt the accuracy of the measurements in the other paper," he remembers. "I looked at a number of other fluorescent solids which might be more suitable. None worked. Each had their own problems.

"I returned to the ruby, trying to understand why the fluorescent process was as inefficient as the paper reported. There were a number of reasons why this could happen, I thought, and looked at all of them, measuring carefully. To my surprise, I discovered the supposed inefficiency did not exist! Instead of having only 1 percent of the green 'exciting' light converted to the red light of the fluorescing ruby, I found that around 70 percent could be converted. Ruby became a real possibility.

"Not everyone agreed with me. I listened to a paper at a conference I attended in which Schawlow, from Bell Labs, said his group had evaluated the ruby and had concluded it was impossible to make ruby work as a laser. I wasn't disturbed by his conclusion. He thought it was impossible. I thought it would be difficult. But I felt strongly that it could be done, even though I was competing with world-class scientists."

Despite qualms by Hughes supervisors as to how successful Maiman's project would be and their suggestion that he switch to work on research into computers, they reluctantly kept allowing Maiman to work on his ruby laser project. Maiman kept going, though he felt the company's financial and psychological support were meager.

"I stuck my neck out a mile," he remembers, "but the concept of being the first person to generate coherent light was exciting! I kept fantasizing about being able to pull it off, being the first in this technical Olympics."

By all Maiman's mathematical calculations, his device should have worked. Step by step, he checked and rechecked his measurements.

His ruby crystal rod was a rod 3/8 inches in diameter and 3/4 inches long. The ends of the ruby cylinder were flat, parallel, and highly polished. He made the ends of the rod reflective by depositing a thin coating of silver on them. He then carved away a small hole in the coating to allow the laser light to escape.

The crystal needed an outside energy source of extremely high intensity to excite the electrons. Maiman researched the characteristics of all known laboratory high-intensity lamps. He considered, then discarded, the idea of using a mercury arc lamp, though his calculations showed it just might work. "If I put my design together and it didn't," Maiman recalls, "I never would have known for sure if I'd failed because I didn't excite the electrons enough or whether I failed because you never could generate coherent light."

Finally, he decided to look at electronic flash. "I went through every catalog of every manufacturer," Maiman says. "I found three lamps listed. They were different sizes but all had approximately the same intensity. To be on the safe side, I sent for samples of all three; but since it was the energy per unit area that counted, I chose the smallest one to try first.

"Usually a lamp like this is mounted in a glass envelope, and its base fits into a socket. I cut off the glass and took off the socket, so I had the bare spiral flashlamp. I mounted the ruby inside the spiral. Around the spiral, I put a very highly polished aluminum reflector. The total housing was just a little smaller than a man's fist."

Maiman theorized that when he turned on the flashlamp, the strobe would put out an extremely intense burst of energy. The ruby crystal, he thought, would absorb that light. The ions would fluoresce and give off red photons. If he could get enough intensity, the red photons would not only glow, but would be amplified.

He predicted that it would take a very short time for light to travel along the length of the crystal and that the amplified light would start to leak out through the small hole he had made in the silver coating. Maiman set up a measuring apparatus, so he'd be able to confirm he had in fact generated coherent light.

Finally, there were no more problems to check out.

Maiman remembers, "I turned it on, and it worked—the first time. When the laser started to 'go', at first the fluorescence was at the low excitation rate. But as we got the excitation higher and higher, the crystal began to act as an amplifier. Then the photons which happened to be going along the axis of the cylindrical rod hit the mirror at one end and were reflected exactly back on themselves to the other mirror!"

Patent Problems

Maiman's achievement in building the first actual working laser was significant, significant enough for him eventually to be named to the National Inventors Hall of Fame, an honor he treasures because there are only about 60 members, despite the 4 million or so patents in existence. Yet even though Maiman had an actual working laser, the first ever built, he had difficulty selling Hughes and scientists on his success and what it might mean.

In fact, the first scientific publication he submitted his work to turned it down. Finally, the British journal *Nature* published Maiman's results. A few weeks before the article appeared, Hughes flew Maiman to New York for a press conference announcing the working laser.

"That was my first encounter with the media," Maiman says. "I described how I thought lasers could be used in medicine and biology, in industry for cutting and welding, and in communications because of the information capacity and enormous possible bandwidth. One reporter asked, 'Is it going to be a weapon?'

"I told him I thought as a practical application of lasers, a weapon is a far-fetched idea. The next day, major headlines in the Los Angeles *Herald* said I'd discovered a science fiction death ray!"

Hughes lost foreign rights to the laser patents, Maiman says, because they didn't file quickly enough. "Later, in 1967, after I'd left and formed my own company to manufacture lasers, I found they weren't processing the patent efficiently. By mistake, I received correspondence between Hughes and the Patent Office saying the patent would be rejected unless a new argument was submitted within 30 days."

As Maiman explains it, the Patent Office traditionally takes an adversarial view. In this case, the office told Maiman that since the ruby laser was "obvious," he shouldn't be entitled to a patent for making one. The office cited previously published papers by other scientists that it said proved their point.

"That infuriated me!" Maiman says.

"I contacted the Patent Office. I told them if they correctly read the references they themselves had quoted and followed the calculations the other scientists had presented, they'd see I was right. The other scientists had said the ruby would *not* work.

"I gave Hughes the chance to use my affidavit, or to relinquish their claim and let me file. Executives at Hughes were delighted to proceed, using my arguments, and the patent was issued within two weeks!"

Even though he'd clearly made the first working laser, Maiman says, it took several years for trade magazines to be convinced lasers actually had value. That taught him a lesson—a lesson he wants to pass on to others.

"If you have an idea you want to pursue, if you've really studied it and thought about it, despite the negative consensus by experts, then go for it! I still remember a class one scientist was to give at a university in the summer of 1960, a class he cancelled after my press conference on lasers. One of his announced topics was, 'Why a laser cannot work.'"

CHAPTER 3

PERSONAL QUALITIES

What personal qualities are necessary to succeed in working with laser technology? What do people with jobs in this field recommend? Here's what several of them say about requirements.

BE ENTHUSIASTIC ABOUT MATH AND SCIENCE

"So much of laser work is related to math and science that you really need to enjoy those fields," says Jack Dyer, communications director of the Laser Institute of America, one of the major laser associations. "The more you understand the optical and mathematical principles behind laser operation, the more chance you have to move ahead in the field, rather than merely running the machinery. You can go much further in your career."

FOLLOW DIRECTIONS PRECISELY

The ability to read and understand directions exactly, and to follow them to the letter every time, is important in laser work. Lasers are unforgiving; if you violate safety standards even once, you risk permanent damage. Various lasers require different types of eyewear. Previous experience with one type of laser does not mean the same safety standards will automatically work with another. In one well-publicized case, a research scientist wearing inappropriate eyewear while trying to look into the path of a Nd:YAG laser received a retinal burn after less than 30 seconds of exposure.

A summary of accidents between 1983 and 1986 reported to the Food and Drug Administration (FDA) by various medical laser equipment manufacturers found that fire, patient reactions, and eye and skin burns were among the problems. R. James Rockwell, Jr., of Rockwell Associates, a Cincinnati, Ohio, firm, says that such accidents can be prevented if laser workers are provided with proper protective eyewear, are properly trained, and follow safety policies.

ACCEPT RESPONSIBILITY

To succeed in laser technology, especially at the technician level, you must be willing to take responsibility. In a manufacturing plant, for instance, you may be one of a handful of laser technicians—or indeed, the only technician—among a large number of other workers or machinists. If the equipment goes down, you've got to get it up and running, no matter what that takes. That may mean trying various "fix" techniques, phoning the manufacturer, or working with plant engineers. The burden of getting the process going again may well rest on your shoulders.

EXPECT TO LEARN ON THE JOB

No matter what your training has been, you will almost certainly find working with lasers a continuous learning experience, says Dyer. "You certainly won't be finished with your training when you graduate!" You'll need to work on a variety of equipment, some of which you will learn about through hands-on experience at the plant or laser site. You'll need to keep up with technical developments through course work, seminars, after-hours study, reading the trades, and similar activities.

BE YOURSELF

"The laser field is broad enough so there's plenty of room for different personality types," says Dr. Frank Jacoby, principal research scientist at Battelle Laboratories in Columbus, Ohio. "If you're the kind of person that likes to sit in the back room and play with equations, there's plenty of opportunity for that; if you're a social sort of person that wants to heal people and interact with patients, the whole medical/laser field is wide open.

"If you're interested in management and the business side of lasers, you can find jobs there, also. So many little laser companies are starting up that you can essentially do almost any kind of work there."

As Jacoby summarizes, there are niches in laser jobs, niches enough to accommodate many types of people. "Laser technology is a much more open field than many others, in which everything is narrowed down and stratified—in which you're either one of 'them' or not," he explains.

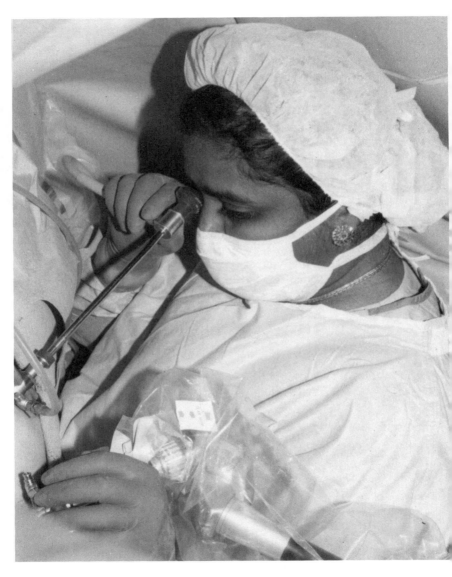

Lasers have become extremely important in health care, and applications in medicine and surgery continue to expand rapidly. (photo by Sharon P. Gregg)

CHAPTER 4

LASERS IN HEALTH CARE

Lasers have become extremely important in health care. Although no one knows precisely how many laser procedures are being performed, the American Society for Laser Medicine and Surgery estimates there are nearly a million such procedures a year. Laser applications in medicine and surgery continue to expand rapidly. In fact, in some medical specialties, according to the society, "virtually no practitioner works without a laser at hand."

The high intensity of laser light and its ability to be precisely focused let physicians control the cutting or cauterizing of living tissues. "You can make a precise cut with very little damage to surrounding tissue," explains Dr. Ellet Drake, the society's executive director. Because lasers can be used without coming in contact with the region of the body that's being treated, there is almost no risk of infection. Because the laser coagulates (seals off) blood vessels as it cuts, the site where the surgery is taking place is almost bloodless, giving the doctor a clear view.

Laser surgery has traditionally worked best on soft tissue, rather than on bone. Because a thermal laser works primarily through heating and cauterizing tissue, the intensely focused laser beam can burn a tiny hole at any desired point with great accuracy. In some cases, a series of these tiny holes are burned into the area surrounding a lesion, helping to seal it off. The procedure is somewhat like spot welding in a factory.

The society estimates that 60 percent of all laser procedures currently being done in the United States involve the eye. Other physicians using lasers extensively include gynecologists, dermatologists and plastic surgeons, pulmonologists, and otolaryngologists. Podiatrists use lasers; so do gastroenterologists and neurosurgeons.

KINDS OF LASERS

In health care, three kinds of lasers are mainly used; the argon laser, the CO_2 laser, and the neodymium-doped yttrium aluminum garnet (Nd:YAG). The laser the physician chooses for a particular procedure depends primarily on where the wavelength of that laser is best absorbed. For instance, the argon laser is effectively absorbed by tissues containing high concentrations of hemoglobin or melanin. Consequently, the argon laser may be the laser of choice for working with tissue where many blood vessels are involved, such as the retina of the eye or a portwine birthmark.

The carbon dioxide (CO_2) laser has a longer wavelength, one that doesn't really differentiate pigmentation. James Johnson, consultant in safety and laser applications and author of two books on lasers, points out that "any tissue that contains water will absorb the CO_2's wavelength. Thus, the carbon dioxide laser is extremely useful for surgery. It can vaporize tissue."

Although the CO_2 laser is used to cut out certain types of cancers, including cervical cancer, it has certain limitations. Doctors using this laser have only "line-of-sight" vision.

Light from the laser can only be transmitted through handheld lenses, through rigid scopes, or through a microscope. The beam of the carbon dioxide laser is transmitted only in a straight line or is reflected by mirrors. Although scientists are working on allowing the carbon dioxide laser beam to pass through flexible fibers, the FDA has not yet approved such a use.

The Nd:YAG laser, on the other hand, can be used with fiber-optic technology. Optical fibers are extremely thin threads of glass that can transmit light over distance with very little loss of intensity. Rays of laser light traveling down such fibers are reflected off the sides.

In certain medical procedures, called endoscopies, the flexible fiber-optic device is used to let doctors look directly into portions of the body that otherwise they could not see, such as bile ducts or the lungs. Endoscopic examinations also allow doctors to treat certain conditions directly.

The Nd:YAG laser beam can pass through such a flexible scope and down to the area needing treatment. The surgeon can direct the laser beam to the target and use the beam to cut out tumors. "The YAG laser penetrates tissue very deeply," Johnson explains, "but its energy is dissipated because it scatters over a large volume of tissue; consequently it's good for cauterizing."

In photodynamic therapy a tunable—that is, an adjustable—dye laser is used. Here, doctors can aim the laser beam at tissue that has been treated with special chemicals. Laser surgery then selectively kills the cancer cells.

Gastoenterologists are using this new procedure on patients with cancer of the esophagus. A multicenter study involving the New York Medical College in Valhalla, New York, and centers in Boston, Columbus, Kansas City, and Miami is evaluating the procedure.

At Valhalla, Dr. Stephen K. Heier reports he and his coworkers have applied the argon dye laser technique eight times to five patients with large tumors that interfered with eating and swallowing. The patients in the Valhalla trials were not suitable candidates for, refused to submit to, or failed at other, older forms of cancer-razing therapy, including radiation and surgery.

In this investigational procedure, a special drug—dihematoporphyrin (DHE)—is injected. The drug is taken up by only the cancer cells.

Light emitted by the tunable argon dye laser is sent down the long, thin, hollow tube, or endoscope, to the portion of the esophagus where the cancer cells are located. Because dihematoporphyrin (DHE) is a photoactive drug, it reacts when exposed to light of a particular color— in this case, the wavelength of the argon dye laser. The drug causes the cancer cells to disintegrate when the pulses of laser light hit them. Because the energy of electrons coming into the laser can be tuned, the resulting laser beam will be just the right wavelength to trigger the process and destroy the cancer cells by stimulating the desired chemical change.

Before this new laser procedure was developed, these patients probably would have been treated with a different sort of laser in a procedure that can damage healthy unshielded cells in the esophagus as well as the cancer cells. The new laser technique, therefore, is much more selective in the tissue damage it causes.

Heier says the drug-and-laser treatment worked successfully to reopen the esophagus to near-normal or normal size. The patients found the procedure to be painless, and they were able to resume eating regular diets. Nearby normal tissue exposed to the laser light received only minimal damage from the procedure.

The new dye-and-laser treatment offers still other economic advantages, Heier says. The older laser costs $100,000 as compared to $40,000 for the argon laser.

The tunable dye laser requires certain additional equipment: another laser to feed in the energy, pumps to circulate the dye, and a cooling system.

Research continues on other kinds of lasers for health care. Excimer lasers are being studied, since they would provide a new range of wavelengths for medical applications, thus affecting tissue differently. Work on these lasers is in the early stages, and they have generally not yet been approved for use on human beings.

Excimer lasers may show promise in angioplasty and in the eye surgery known as radial keratotomy. Free electron lasers are also being studied.

LASERS IN OPHTHALMOLOGY

In the early 1960s, soon after lasers had been invented, doctors and scientists began experiments to see if the technology could be used to treat eye disorders. By 1961 a group headed by Dr. Charles J. Campbell at Columbia Presbyterian Hospital in New York City became the first to use the laser to treat a detached retina. Several years later, Dr. Hugh Beckman used a laser to open tiny holes in the iris. This treatment, performed at Detroit's Sinai Hospital, demonstrated that lasers could be used successfully as a surgical instrument as well as in coagulation.

Today, this operation is extensively used to treat narrow-angle glaucoma. The tiny holes the laser makes in the iris let fluids flow easily again, thereby reducing the dangerous buildup of pressure caused when obstructions block the normal flow of fluid from the main body of the eye.

The argon ion laser, which uses a medium of electrically charged argon gas for generating laser light, is used extensively in eye treatment—for a special reason. The argon ion laser gives off a blue-green light, a color that's strongly absorbed by red objects, such as blood. Many eye procedures involve sealing off blood vessels, so this laser becomes a powerful tool in coagulation.

Diabetic retinopathy is an extremely common complication of diabetes in which abnormal blood vessels develop in the retina. With the argon ion laser, the ophthalmologist can release up to 3,000 laser bursts that strike the retina in a predetermined pattern. These short bursts of light can slow or stop bleeding from the blood vessels.

Argon ion lasers are also used to treat macula degeneration—an eye condition that often develops when a person becomes old. The macula is the central portion of the retina. If it becomes diseased, fluids can leak out from behind it and hamper vision. Because the laser coagulates, it can help reduce the leaking.

Lasers are also used to treat another form of glaucoma in which pressure buildup cannot be traced to a specific abnormality. The ophthalmologist uses the laser to make a series of openings in the tissues that normally let fluid flow out from the front chamber of the eye.

Ophthalmologists use the argon ion laser a great deal. In addition, they also use a krypton laser, which is similar in design. It uses krypton, another gas, as the lasing medium. Ophthalmologists choose these lasers because of their photocoagulation ability. Although the argon ion and krypton lasers use an extremely small amount of heat, that heat can be delivered very precisely to the area the beam hits.

In addition, ophthalmologists like Dr. George Sanborn, associate professor of ophthalmology at Southwestern Medical School, University of Texas Health Science Center, Dallas, use another type of laser—the Nd:YAG laser—to treat certain types of glaucoma. They also use it to treat a condition in which patients who have had previous cataract surgery and who have had an artificial lens implanted develop clouding in a normally clear membrane behind the lens.

"The YAG does not use heat," Sanborn explains. "It's a disruptive instrument. It vaporizes the tissues it hits." Rapid pulses of the Nd:YAG laser eliminate much of the clouded area of the membrane, yet they keep enough of it intact to hold the artificial lens implant in place.

Still another type of laser, the excimer laser, is being tested for use in a controversial procedure called radial keratotomy. Some ophthalmologists are using radial keratotomy to make many tiny incisions in the cornea in an attempt to improve nearsightedness. Scientists who are studying excimer lasers think that perhaps the lasers can make fewer and shallower cuts in the cornea than the cuts needed when a surgical scalpel is used.

The carbon dioxide (CO_2) laser used in general surgery won't work for ophthalmology, Sanborn says, because vaporizing the tissue inside the eye would fill the eye with gas bubbles.

"Ophthalmologists today regard laser surgery as almost routine and not at all unusual," Sanborn says. "Today, most opthalmologists receive training in laser surgery as part of their residency. Generally, they spend two months on the 'retina service' during their second year of residency, and that is where they're introduced to lasers in a structured program. By their third year of residency, they're using lasers (under supervision) to treat patients."

LASERS IN OBSTETRICS/GYNECOLOGY

Reproductive endocrinologists like Dr. Reena Jabamoni, a fertility specialist, are using lasers to help treat conditions that may interfere with conception. "The primary laser I've used is the CO_2 laser," she explains. "I use it for pelvic endometriosis, for reconstruction of the Fallopian tubes that have previously been damaged by endometriosis or pelvic inflammatory disease, and for removal of fibroids, which are benign tumors of the uterus."

Another type of laser, the endoscopic Nd:YAG, can be used to treat excessive menstrual bleeding not caused by an underlying disease. "I use the Nd:YAG laser to destroy the lining of the womb—a procedure which helps many women to avoid having a hysterectomy," Jabamoni explains. "However, this procedure makes a woman infertile, so it is used only when a woman plans not to have children."

One of the reasons for using the CO_2 laser in gynecology is that abnormal tissue can be vaporized or cut out easily. Gynecologists use the laser to remove lesions of the cervix, vulva, and vagina. Since much of a woman's reproductive system is accessible from outside the body, they can also use the CO_2 laser to cut out a tiny cone-shaped section of the cervix for diagnosis. Such "laser conization" can be done quickly and with virtually no bleeding. Women who do want to have children have very little risk of losing their fertility when this procedure is used.

Lasers are also used by gynecologists to treat other conditions. Genital warts, caused by a virus, can be removed with the laser—and the rate at which the warts seem to reoccur is substantially less than it would be with other forms of treatment. A laser inserted through a small incision in the abdomen can be used to treat adhesions of the ovary— a condition that affects fertility and occurs when the outside of the colon or another structure in the pelvis becomes attached to the surface of the ovary.

Reconstructive surgery of the uterus is another area in which gynecologists can use lasers. "In a rare condition when a woman has a uterus divided into two compartments, I use the laser to take out the tissue between the compartments. Then the uterus becomes one full-size cavity that can hold a baby," says Jabamoni.

Procedures like laser laparoscopy are performed as outpatient ambulatory surgery, she says, while she has patients stay in the hospital when she is performing more extensive reconstructive laser surgery.

Although 40 percent of her practice today uses lasers, Jabamoni says that she started out in 1979, before the use of lasers in gynecology "took off." Consequently, she received her laser training after setting up her practice.

"First, you take classes," she says. "Then you have hands-on experience in a laboratory, first in inanimate objects, and then with animal experiments. Next, you work with a preceptor—someone who has very good training. After the preceptorship, you will probably do a number of cases in a hospital setting, under the supervision of an experienced laser surgeon." Although the laser is precise, she says, "when you aim the beam, you are destroying tissue, layer by layer. Because you are working close to important organs such as bladder, ureter, and bowel, in vital areas of the body, you must be extremely controlled in your technique."

Keeping up with new developments in laser use is "mandatory," Jabamoni feels. She herself belongs to the Gynecological Laser Society and the American Society of Laser Medicine and Surgery and attends meetings of both organizations regularly.

LASERS IN PODIATRY

Podiatrists are using lasers more and more in their practices. Just how much more, though, depends on the type of practice a podiatrist has. Fortuneé Massuda, D.P.M., a Chicago podiatrist, estimates that only 5 percent of her work involves laser surgery. "That's because I see a lot of patients with bunions and hammer toes," she explains. "I don't use lasers for bone surgery, because I don't want to risk burning the bone."

Laser surgery is useful, she says, for soft tissue conditions: warts, plugged sweat glands, ingrown nails, and corns that are not due to a bone deformity. According to Massuda, laser surgery for these problems is simple, quick, and relatively painless compared to conventional surgery. "Patients aren't frightened by laser surgery," she says. "They're excited about it. They think it is magic."

In 1984, when Massuda bought the CO_2 laser she uses in her office, the price was $35,000. By 1988 machines had become smaller in size, and the price had dropped to around $27,000, which in Massuda's view was still expensive.

From the patient's point of view, laser surgery is not difficult or prolonged. "My associate anesthetizes the area," Massuda says. "I go in, use the laser to cut out the diseased tissue and coagulate blood vessels. I complete the surgery quickly. For instance, I can remove one wart in 2 to 3 minutes—or a group of warts in less than 10 minutes. My associate dresses the wound, and the patient leaves the office."

One important precaution: safety glasses. Massuda, the patient, and anyone else in the room wear them during the surgery because of the intensity and penetration power of the laser beam.

Since Massuda was already practicing at the time laser surgery for podiatry became common, she received her initial laser training in seminars run by the company that sold her the laser. "That's the usual way," she says. "The vendor wants to be sure you are using the laser properly. You also want to be sure you have proper certification showing you know how to use lasers, in case you're ever sued for malpractice."

Today laser companies often donate lasers to podiatry schools, so students can be trained in laser surgery. "It's good business for the companies," she says. "In addition to good will, they know that students who have done this surgery as part of their podiatric training will be comfortable when they use the laser. When they go into practice, they'll want access to a laser—either in their own offices, or in a hospital or free-standing surgicenter."

WORKING WITH LASERS

Although many doctors use lasers routinely in their practice, not everyone who works with lasers in health care is a physician. Laser nurses play an important role in hospital or clinic care. Sometimes these nurses are also designated as laser safety officers; at other times, the laser safety officer serves only in that post and does not function as a nurse.

A LASER NURSE COORDINATOR

At Wenske Laser Center, Ravenswood Hospital, Chicago, Robert Pyrcz serves as laser nurse coordinator and as laser safety officer, directing a staff of three. A graduate of Northeastern Illinois University in secondary education and U.S. history, Pyrcz spent 10 years teaching mathematics to eighth graders in Chicago public schools before deciding to make a career switch. Although he enjoyed teaching, he didn't like being transferred to a new school every few years. He wanted to work with people but found himself spending more and more time on paperwork.

Checking want ads to see where jobs were available, Robert found only one or two columns of classified ads for teaching positions— but four or five pages worth for nurses. Other family members who were nurses liked their careers and talked about them often; Robert thought he too would enjoy nursing.

He enrolled in Ravenswood Hospital's School of Nursing in 1981 and graduated in 1983. Because he'd had a number of science courses in his undergraduate studies, he could shorten the study time usually required for the nursing curriculum. He passed the state examinations and became licensed as a registered nurse (R.N.).

After a year's experience at Ravenswood Hospital in the rehabilitation unit, Robert saw an opening posted for a laser nurse. He interviewed with the hospital's medical director and was offered the position. Initially the hospital sent him to an intensive three-day workshop in Columbus, Ohio; after that, Robert says, it was basically "on-the-job" training, learning as much as he could from the physicians and by being around lasers.

Ravenswood Hospital (a major Midwest laser center) has 10 lasers: 2 tunable dye lasers, used primarily in research; 2 argon lasers, used for eye procedures and for dermatology; 1 Nd:YAG laser, used for general surgery and gastroenterology; and 5 CO_2 lasers, used by physicians in various medical specialties.

Pyrcz divides his time between the laser center and the regular operating room of the hospital. Patients who can have laser treatment using a local anesthetic or no anesthetic at all (such as eye cases or cases involving the removal of warts from feet) are treated in the center.

Patients whose surgery requires general anesthesia receive laser treatment in the operating room.

Patient education is an important part of Robert's job. He visits patients before the operation, letting them know what to expect with laser surgery. "Many times, they're a little afraid of the laser," he says. "They want to be reassured that it won't drill holes into them. I explain that laser surgery is extremely precise and talk with them about how it will help their condition."

Pyrcz assists surgeons during the laser part of an operation. He conducts workshops for nursing personnel, physicians, and biomedical technicians—a responsibility that takes up at least 30 percent of his time. Another 20 percent, he estimates, goes into public relations, such as speaking to various community groups. Half his time is left for patient care (including surgery) and patient education.

A Typical Day

On a typical day Pyrcz arrives at work at 6:30 A.M. His first task is to check the surgical schedule to see what types of operations are going to be performed with the lasers.

He describes a day on which he found there were six laser operations scheduled in this way: "The first one was hemorrhoid surgery, with the CO_2 laser. Then there were two eye cases back-to-back. One was eye surgery to correct a condition that occasionally occurs after cataract surgery. We planned to use the Nd:YAG laser. The second was a panretinal photocoagulation to get rid of abnormal blood vessels that grow in the retina as a complication of diabetes—a procedure in which we would use the argon laser. Finally, there were three cases in a row in which we were destroying bladder tumors with a high-powered Nd:YAG laser."

For safety and security (since some of the lasers have cost $100,000), treatment rooms containing them are kept locked. Robert, who describes himself as "the keeper of the keys," unlocks the rooms to prepare them for the surgeries. Additionally, each laser is locked with a separate key, which Robert carries.

"I wheel the laser into position," he says. "I prepare the operating table by draping it, and open up the instrument trays.

"I plug the laser in on its special electrical circuit, since it takes an immense amount of power. I also wheel in the smoke evacuator and set it up, to take the vapor caused by the impact of the laser beam on tissue out of the air."

Robert also makes sure the medical records for each patient are set out on the table, ready for the doctor. Careful, detailed records are kept by the hospital's medical records department and are used for analysis and reference.

For each procedure the physician lists the power settings, the type of laser used, the amount of time the patient received laser treatment, how the laser beam was delivered to the tissue, the safety precautions taken to protect the patient, the verification of the patient's consent to operate, the type of operation performed, and the diagnosis of the patient's condition. As laser nurse coordinator, Pyrcz checks and keeps a copy of all this information.

Later, records will be analyzed so the hospital can determine general trends in laser surgery: What kinds of power settings are doctors using? What seems to be the standard mode of operation in a particular type of laser procedure?

Robert is present at all laser surgeries, both in treatment rooms and in the hospital's operating room. In the latter he is scrubbed, gowned, masked, and gloved. After the patient has been put under general anesthesia, Robert prepares the laser for use in a sterile field. He puts sterile drapes (or cloths) over the arm of the laser. He drapes the smoke evacuator to keep it sterile.

The physician announces what power setting he or she has chosen for the treatment and whether he or she wants the laser beam delivered in a continuous wave or as short pulses. Robert adjusts the laser accordingly, making sure it is set properly and functioning correctly.

During surgery, as laser safety officer, he also has the responsibility of checking that safety procedures are being followed. "Everyone— including the patient—must wear proper eye protection at all times," he says. "The patient's operative site must be covered with wet towels to protect the surrounding skin that isn't being treated. I make sure the doors to the operating room are shut, that warning signs are posted showing laser surgery is going on, and that no one comes in or goes out. I make sure there is always water available, if a fire should occur."

If anything in the room, or during the laser surgery, does not meet Pyrcz's standards or satisfaction, he has the authority to stop the treatment and turn off the laser.

Even though normal working hours for Pyrcz are 6:30 A.M. to 3:00 P.M. Mondays through Fridays, with weekends off, he is on call for emergencies and has to wear a beeper. "I get beeped all the time when I'm in the hospital," he says. "Usually I can deal with the problem by phone. About four times a year I get calls at home. Sometimes I can handle the situation over the phone, but I'll generally get dressed and go down."

The salary for his position, says Pyrcz, is approximately $30,000. The three staff members he supervises (all of whom he trained) perform similar duties and receive approximately $24,000.

For a position like his, he says, a man or woman first must qualify as a registered nurse. "Then, if you have the interest, there are many workshops and seminars available so you can get the necessary training and continuing education hours you need to meet standards in your hospital."

A LASER COORDINATOR

Holding a position at Rose Medical Center in Denver that is similar to Bob Pyrcz's is Susan Hicks. While studying sociology at the University of South Dakota, from which she graduated with a bachelor's degree in 1976, she also took courses sufficient to earn an associate degree in nursing. For a time she worked as a registered nurse (R.N.) in South Dakota.

A move to Denver gave her nursing experience in liver transplants and intensive care. She didn't enjoy rotating between day and night shifts, however, and looking for stable hours, she moved to Rose Medical Center. After three years' experience nursing in the recovery room, she helped set up the center's first outpatient surgical area.

"The center bought its first laser in 1983," Susan remembers. "Since I was the most interested, I volunteered to take courses and learn more about it. I wanted to get credentialed in laser safety and to help train other nurses."

Administratively, Susan's present position as laser coordinator combines many responsibilities of a laser safety officer with those of a clinical coordinator. She helps educate nurses on laser procedures. In addition, she is at the center for every laser surgery performed, making sure the equipment is properly set up and all safety features are in place.

In March 1988, she says, the center performed 115 laser surgeries. Rose Medical Center has an Nd:YAG laser and an argon laser for ophthalmology procedures, an argon laser for gynecological and dermatology surgery, two CO_2 lasers for general surgery, and a continuous wave YAG laser with fiber-optic capabilities that's used for pulmonary, gynecological, and gastroenterological treatment.

As she prepares operating rooms for surgery and checks laser safety features, much of Susan's work is similar to that of Bob Pyrcz's. She is also deeply involved in a center task force that is investigating how to start a laser program for vascular procedures. "I spend a lot of time with vendors," she says, "because as soon as anyone hears you have a laser program, they want to sell you something. I spend much of my time doing physician marketing, explaining the lasers and helping them (physicians) arrange a trial run, setting up a practice session for them with the equipment. I troubleshoot when there are laser problems or problems with related equipment, working closely with Bill Van Dyken, our biomedical technician. The laser is a temperamental machine, and it's good to have a backup person in the hospital. If I have to take the laser apart, forget it! Bill has a background of a medical engineer, and can do it."

Since she comes to work at 7:30 A.M. and often doesn't leave until 6:30 or 7:00 P.M., it's a long day for Susan. Rose Medical Center has just hired an assistant for her, which will ease the pressure.

Getting out from under some of the responsibility will free Susan to complete other projects like the one she's started on eye testing for staff personnel. Certain wavelengths, like that of the argon laser, can penetrate right through the eye to the retina and can damage patients' or staff members' eyes if a stray beam escapes or reflects. Susan is making sure all staff who work with the lasers have received eye tests. Results are recorded in their personnel files. If they leave the center, their eyes will be tested again and records compared to be sure they haven't sustained any eye damage.

"Every day, I'm still finding out what my job is," Susan says. "It's a lot of hard work—but it's really rewarding."

LASER SAFETY OFFICERS

Not all hospitals have full-time laser safety operators. At Grant Hospital in Chicago, Frank Hurley, trained as a clinical perfusionist, carries that title. But, says hospital administrator Jay Tuke, Frank may spend only one-third of his time on laser issues.

"Once things have been set up and are running smoothly," Tuke says, "Frank's time as a laser safety officer isn't that much. He does have to know how to operate the equipment, but for much of his day, he works as a clinical perfusionist."

Grant Hospital began offering laser surgery in the fall of 1987. A multidisciplinary laser committee considers and writes policies and procedures that Frank has the responsibility of enforcing. Sitting on the commitee are Frank, Tuke, a representative from obstetrics/gynecology and one from gastroenterology, a biomedical engineering representative, and an administrator to whom operating room personnel report. The chairman of the interdisciplinary laser committee is the chairman of the department of surgery.

Grant Hospital uses its Nd:YAG laser for general surgery, such as mastectomies, thyroidectomies, hemorrhoidectomies, and some gastroenterological procedures. The hospital's two CO_2 lasers are primarily used for gynecologic procedures, often for surgery related to infertility.

Although Grant Hospital hopes to expand its laser surgery, it doesn't foresee hiring additional personnel who haven't worked in hospitals. "I think we'd take a staff member with experience as a hospital operating room technician or anesthesia technician," Tuke says. "Then we'd teach that person laser skills."

"Laser is a sexy term, but when you get down to it, lasers are just another medical tool. To use lasers in medical applications, you have to be trained first in the medical field. Then you add laser training to your basic knowledge."

LASER EDUCATORS

Although some hospitals and medical centers are teaching laser courses to their own staff members, others bring in outside consultants to perform that function. Colorado-based Education Design is a company that puts on such courses for national and international clients.

President Judith Pfister says these courses can range from a three-page self-learning booklet on a technique, product, or procedure; to a one-day workshop; to a textbook or training manual; or to a full five-day convention. Judith also chairs the nursing section of the American Society for Lasers in Surgery and Medicine, helping to design in-service courses given at meetings and conventions.

A graduate of a three-year nursing school, Pfister also holds a bachelor's degree. She first was introduced to lasers in 1968. As she progressed up the career ladder in nursing, eventually becoming assistant director of all the suites of operating rooms for Methodist Hospital in Houston, Judith was introduced to a variety of lasers, as various doctors in the facility became interested in that technology. Several years of work as assistant executive director of the Association of Operating Room Nurses (a national professional organization) taught her management techniques. In 1981 she and three other nurses founded and incorporated Education Design. Their target market: hospital surgicenters, clinics, facilities, and hospices that need laser education—and the medical industry's manufacturers and distributors of laser health care products. She sees her company as providing training materials and personnel suitable for both.

Carol Wirth, R.N., another laser educator, is president of Leadership Concepts Incorporated. The Wisconsin-based firm deals specifically with lasers in health care and operating room management. "We help hospitals look at their program," Carol says. "If they don't have a laser program, we will do feasibility studies. We made recommendations for what type of program they should be formulating. We give them a program description, write policies, procedures, and job descriptions, and help them implement the program." Leadership Concepts Incorporated presents hands-on, two-day workshops that help nursing staff members at the facilities understand lasers and laser technology.

In addition, the company acts as an all-round resource for clients on an ongoing basis, keeping them aware of changes in technology.

A graduate of a nursing school in Janesville, Wisconsin, Carol worked her way up from a staff nurse at Mercy Hospital there to operating room supervisor. Recruited by a Rockford, Illinois, hospital, she spent five years there, helping to develop programs to train staff personnel. At Rockford she saw firsthand the importance of laser training when the hospital purchased its Nd:YAG and CO_2 lasers. She had worked well with independent consultant Carolyn McIntee, who headed Laser Consultants, Inc., and the two joined forces in 1984 to form the new company.

Hospitals are eager for training courses in laser operation and laser safety, Carol says—not only because they want to reduce their liability by being able to prove staff members have been properly instructed, but also because interest is growing. The prime benefit hospitals receive from these courses, she feels, is having trained and educated personnel to run these sophisticated, yet hazardous, pieces of equipment.

Is a health care career working with lasers for you? Yes, say practitioners, if you like the medical environment, are willing to work hard, and are interested in keeping up with technological developments. Medical research is moving so quickly that continuing your education is a "must" in this demanding profession.

A heat treat specialist is trained to use a CO_2 laser for selective hardening of gear tooth surfaces. (photo courtesy of Garrett Engine Division, Phoenix, Arizona)

CHAPTER 5

LASERS IN MANUFACTURING

The special properties of laser light (monochromaticity, coherence, divergence, and brightness) make it useful in manufacturing. It is so useful, in fact, that lasers have become important in many industrial processes. Lasers play a significant role in drilling, cutting, marking, welding, heat treating, and cladding (a technique used to melt alloys and selectively deposit them onto surfaces of parts).

Many people who imagine themselves having a career with lasers in manufacturing picture themselves as running the laser. In today's factories, however, a laser is usually not a "stand-alone" machine. Instead, it is part of an automated process—a process that involves a number of steps.

A manufacturing system that uses lasers needs a way of delivering the part being handled to the workstation where the laser operation will take place and of taking the part away to the next stop after the laser process has been completed; a laser system capable of generating sufficient energy to perform the desired task; and a way of focusing the laser beam that delivers the energy to the "right" spot on the part (a task often handled by automated vision systems). Because only part of the electrical energy can be used as optical output (the laser beam), a laser manufacturing system needs a cooling system to get rid of the rest of the energy and resulting heat. If the laser process involves removing or vaporizing material (such as in drilling or cutting operations), an exhaust system is needed to remove smoke, gas, and particles of material from the workstation.

In addition, a manufacturing system involving lasers almost certainly will have a controller to make sure everything is happening for the "right" length of time and in the "right" sequence. It will probably also have an off-line programming system, so that changes in the computer program governing the operation can take place without having to shut the assembly process down. Although industrial lasers have been in use for more than 20 years, their popularity is increasing as more and more factories begin automating. Initially, vendors supplied turnkey systems. Even today, the capital investment required for installing a large, powerful laser system is significant—often as high as $150,000 to $500,000. However, technology has dropped the cost for certain kinds of lasers. Consequently, smaller factories are finding lasers affordable and are beginning to install one or two of them for "job shop" work.

WHY LASERS ARE USED

There are several important reasons why lasers are economically worthwhile for many manufacturing processes, despite the high initial cost and the length of time it takes to train workers to use lasers properly.

In such traditional manufacturing processes as welding, cutting, and drilling, tools come in contact with the parts being worked on. Friction and the abrasive contact between tool and part wear out machine tools, so they must be replaced often. Tool replacement is much less frequent with lasers, because the beam is performing machining operations without direct contact.

Since the laser beam can be accurately and precisely focused easily, the energy needed to accomplish the work can be placed accurately. Very little of the material surrounding the site of the laser weld or laser cut is affected by the heat resulting from the energy that's delivered. Consequently, the material being worked on has less damage and waste. Parts are less deformed. That's an important advantage when work is being performed on parts for an aircraft engine or when lasers are being used in small welded assembly jobs.

Another advantage of laser processing is its flexibility. It is possible to control the heat and energy of the laser by changing its settings to give the beam more or less power as desired.

Speed is another important advantage that lasers have. In many cases lasers can process materials at a substantially faster rate than more traditional methods. The line, or job, moves faster—thus saving money.

Laser tooling also has the advantage of being very simple. Most lasers have multiaxis capability—that is, they can be positioned to work in different directions. If a part is round, or if laser work (such as a weld) must be done at a difficult angle, the laser tool can usually handle the job. Because the laser beam can be focused to a very small diameter, lasers can be used to work on areas of parts that otherwise would be hard to reach.

TYPES OF LASERS

Lasers used to process material are generally of two types: solid-state or gas. A solid-state laser uses a crystal for the lasing medium. A gas laser, such as the carbon dioxide laser (CO_2 laser) uses the gas as the lasing medium.

Some lasers, such as the neodymium-doped yttrium aluminum garnet laser (Nd:YAG laser) and the CO_2 laser, can be operated to produce a continuous wave (CW). Others, such as the ruby laser and neodymium-doped-glass laser (Nd:glass laser) are set up to provide short pulses of laser light. The Nd:YAG and CO_2 lasers can also be operated as pulsed lasers, if such a mode is appropriate for the desired material handling process. The solid-state alexandrite laser is another such laser that can operate either in a pulsed mode or as a continuous wave.

The choice of a particular type of laser to use depends on a number of factors, including the power range needed to perform the desired task. For example, YAG lasers are exceptionally well suited for precision drilling and cutting operations on a wide variety of inorganic materials where power ranges of 0.5 kilowatts or less are required. YAG lasers have two important advantages over conventional tools in these applications. The laser beam doesn't get dull or change its size or shape as a result of wear. In addition, the way the laser beam performs is not substantially affected by how hard or how machinable the piece is that the laser is working on.

CO_2 lasers are the powerhouses of industry. They're able to operate at power levels up to 10 kilowatts—often making them the technology of choice for heavy-duty industrial applications like cutting, welding, and heat treating. The CO_2 laser is also well suited for processing organic materials.

For applications needing low power, the YAG laser is generally chosen; for high-power requirements, it's the CO_2. In between, there's a narrow range of power requirements where either the YAG or CO_2 can be used.

YAG laser systems have successfully drilled a wide variety of high-temperature alloy aircraft turbine components, automotive gear and bearing materials, and other precision parts. In addition, they are used for cutting complex shapes in products ranging from steel automobile doors to titanium aircraft components, and for welding a variety of industrial products.

Carbon dioxide laser applications include cutting and welding a variety of metals, up to 25 millimeters thick, cutting, drilling, and welding plastics and such other organic materials as leather, wood, and paper; cutting and drilling glass, quartz, ceramics, and similar materials; and heat treating and surface treating of a variety of metals. If desired, a CO_2 laser can be attached to a robot manipulator and used for production welding, cutting, heat treating, and similar applications.

If a manufacturer needs flexibility (and if it's appropriate to do so), a YAG laser and CO_2 laser can be combined on the same laser processing system. Both the YAG laser and the CO_2 units can be permanently attached and can share the same delivery system for the laser beam. Or if preferred, both laser units can be configured and installed as stand-alone units, so that the YAG laser and CO_2 laser have individual delivery systems.

Vendors work closely with factory engineers and process planners to match laser characteristics to application requirements in order to achieve maximum efficiency in a highly cost-effective installation.

For instance, Raycon Corporation, a leading vendor of laser

processing systems, reports that a manufacturer of jet engines needed a laser machine tool that would work well with many different aerospace metals, different-shaped parts that were being processed, parts that varied in size, and different processing applications. There was little room for error in finished parts; the machining tolerances were demanding. In addition, the parts being worked on had to be placed and held in an extremely accurate position so the laser beam could be focused at just the "right" point.

Raycon worked closely with the manufacturer to develop a general-purpose laser processing system that was both versatile and precise—while still meeting the particular application requirements. Although similar systems can be installed for other manufacturers, they too can be modified, either with commercially available equipment or with custom-made hardware. For example, a CO_2 or Nd:YAG laser can be equipped with a special nozzle assembly, with output power that's matched to a particular job, or even with a completely different machine control system.

MORE SOURCES OF INFORMATION

Finding out more about how lasers are used in manufacturing is easy. Many of the books listed in appendix B have information on various laser applications. An excellent overview that's technical, but still easily understood, has been written by Gary F. Benedict in *Nontraditional Manufacturing Processes.*

Benedict, senior project leader at Allied-Signal Aerospace, Garrett Turbine Engine Company Division, has devoted one chapter to laser uses.

Current technical articles about "Lasers in Manufacturing" are indexed in *Applied Science and Technology Index,* found on reference shelves in almost every public library. Many of the magazines cited are listed in this book's appendix B, along with information on how to obtain copies or begin a subscription.

An excellent source for papers on lasers in manufacturing is the Publications Division of the Society of Manufacturing Engineers, One

SME Drive, Dearborn, MI 48121. The society's telephone number is (313) 271-1500. SME's reference librarian can help you locate particular papers by running an on-line search for titles and abstracts (check first, since there's a fee). Or you can find the titles of papers you want by checking the current edition of SME's publications catalog. Then you can order reprints of the papers you want. They cost $4.50 each for non-SME members; $3.75 for members.

A set of seven laser technical papers (#1313-0845) can be ordered by writing to SME's Publications Sales, One SME Drive, PO Box 930, Dearborn, MI 48121 or by calling (313) 271-1500, x 418 or x 419. The cost of the set is $21.00 for SME members and $25.00 for nonmembers.

Such papers as "Laser Selection for Drilling" and "Laser Selection for Cutting" (SME #MRR86-312 and #MR87-235) give specific information about particular laser applications. "Lasers in the Future" (SME Paper #MS86-309) discusses growth trends for lasers that extend the capabilities of flexible machining centers, as well as for heat treatment by laser. SME paper #MS86-314 lists key questions managers should ask before choosing a laser as the preferred manufacturing technology.

Reading papers like these will help you learn more about lasers in manufacturing and will give you a chance to see if a possible career in that field appeals to you.

Meetings and Conferences

Another way to learn more about lasers in manufacturing is to attend meetings at which they are discussed. On the local level, chapters of SME's various associations—especially the Computer and Automated Systems (CASA) and Robotics International (RI) divisions—hold dinner meetings at which speakers present various topics. Such meetings are usually publicized, and nonmembers can attend by registering in advance and paying a small fee.

Trade shows, such as SPOT (sponsored by SME) and DE-LASE (sponsored by SPIE—the International Society for Optical Engineering), are excellent sources of information. You can read more about these in appendix A. Smaller clinics, such as "Lasers in Electronics Manufacturing" (SME-sponsored), are focused on particular topics. Papers presented at these clinics often provide case studies of laser systems at individual firms ("Microelectronic Laser Welding—Proven on the Production Floor") or offer discussions of technical and economic aspects of laser work ("Process in Laser Marking and Engraving").

Films and Videos

Another way of learning about lasers in manufacturing is to look at videotapes. SME's 1988 release, *Lasers in Manufacturing,* is a 35-minute tape about fundamental operating principles, common benefits, and typical laser equipment available. Four case studies demonstrate how using lasers can improve manufacturing productivity and profitability.

At Allen-Bradley, laser gauging and marking of contactors and relays gives increased flexibility. At Cardiac Pacemakers, lasers trim impulse generators, reducing the possibility of damage to fragile parts. At General Motors Saginaw Division, lasers treat rear-axle tubes for uniform hardening and minimal distortion. And at Harley-Davidson, lasers trim motorcycle parts, reduce inventory requirements, and cut changeover times.

In addition, the tape features four industrial experts: Stanley Ream (Battelle Columbus), Richard Scherer (Photon Sources), Robert Schmidt (Laserdyne), and Frederic Seaman (formerly of Illinois Institute of Technology Research and now head of F.D. Seaman & Associates.) They share their experiences and perspectives on lasers in manufacturing and talk about how lasers can combine with other machine tools to increase productivity.

SME also has videotapes that can be rented or borrowed by schools. Well over a million viewers in school and community audiences have seen the career film, *The Challenge of Manufacturing,* available free of charge to schools, career counselors, and libraries, which encourages young people to relate their personal interests and talents to a manufacturing career. The 24-minute tape does not specifically target working with lasers as a career; however, it does show the factory environment, and it does discuss activities young people can do while they are still in high school to help prepare them for manufacturing engineering.

Contact the SME Education Department, One SME Drive, Dearborn, MI 48121 for purchase or free-loan information. Lending is also handled by Modern Talking Pictures Service, Inc., Film Scheduling, 5000 Park Street North, Saint Petersburg, FL 33709. *The Challenge of Manufacturing* is available either as a 16mm film or a videocassette. Teaching materials also are available for use with *The Challenge.*

Another educational film about manufacturing, released in late 1987, is *Race Against Time.* This twenty-eight-minute film from SME is aimed at college-level students and adult audiences. It focuses on trends and new opportunities in manufacturing, competition, automation and the computer, the teamwork concept, and the rewards of manufacturing.

Examples from IBM, Briggs & Stratton, Intel, McDonnell Douglas, Ingersoll Milling Machine, and Seafreeze are shown. Also featured is commentary from prominent experts, including Dr. Tom Peters, author and consultant for business excellence.

Race Against Time is available both as a 16mm film and as a VHS videocassette on a free-loan basis from Modern Talking Picture Service, Inc., 5000 Park Street North, Saint Petersburg, FL 33709 and from SME, One SME Drive, PO Box 930, Dearborn, MI 48121.

KINDS OF JOBS

There are several levels of jobs available for people who want to work with lasers in manufacturing, says Fred Seaman, laser veteran who heads F.D. Seaman and Associates and who spent over 10 years running the laser center at the Illinois Institute of Technology. There's certainly a need for people who are "doers" and who have only a limited knowledge about lasers. Often men and women who actually operate lasers in manufacturing come from other backgrounds. Perhaps they have been numerical control (NC) operators with previous experience in running controllers that oversee automated operations. With previous experience in machining or in tool and die work, Seaman says, they can begin to operate lasers after relatively minimal training.

Training for these people is relatively simple, Seaman feels, and can be carried out by complex interactive graphics. At Ford Motor Company, Seaman says, a worker receiving such training watches a computer screen on which questions appear. If the worker gives the "right" answer, he or she continues on to the next question; if not, the screen displays an animated character who indicates the mistake.

Salaries at this level average approximately $9 per hour in direct take-home pay, $8 or $9 per hour in a job shop, and possibly $10 to $11 an hour in the automotive industry.

The next level up from operations is "maintenance," Seaman points out. Here, he feels, a worker needs at least a two-year educational program about laser principles. "He or she has to know enough about lasers not to get hurt, as well as enough to understand the purpose of what's being done with the laser," Seaman says. The maintenance worker will clean portions of the laser system periodically and replace worn parts. Usually the duties are limited to following a checklist, doing the items as predetermined. For example, one task might be "take out the lens and clean it." Another might be "check filter."

"After he takes a lens out, he has to understand what test he will use to see if he replaced it correctly," Seaman explains. "He or she may have to do some things with cooling systems, so the person must know a little about refrigeration. Most of the knowledge needed is taught at the two-year-college level, in technicians' programs."

Maintenance people may receive $13 or $14 per hour and are likely to belong to a union.

Service personnel—a higher level than maintenance people—are generally employed by vendors and visit customer sites to fix problems or troubleshoot the laser system.

"They're detectives," Seaman says. "Although they listen to what operators and maintenance people tell them about what has happened and why the laser doesn't seem to work properly, a service person has to decide whether those people are really good observers or are merely passing on their preconceived ideas. Service persons need to work well under pressure and time deadlines. They're under the gun from plant management. And sometimes they have to stand before the company's board of directors, explaining just why the expensive laser system isn't working."

Seaman says service people need a real ability to solve mysteries. In addition, they do their own observations of manufacturing processes, seeing what problems are happening and trying to clear them up. In order to identify problems, causes, and remedies, service personnel must know how to use typical electronic diagnostic tools, such as meters and scopes, and analytical optical equipment.

Salaries for service personnel run about $20,000 to $25,000 per year. Extensive travel may be involved, because service is performed at the client's plant, wherever it is located.

Although there are no technical reasons why women cannot be service personnel, given knowledge and experience, "it's an old boys' club," Seaman says. "The hours are long, and there's significant physical effort. When those guys get in a manufacturing plant, they are often having to lift heavy equipment. They're usually pretty husky. It's a tough life."

Above service personnel in this informal hierarchy, Seaman explains, are applications engineers. Often these are service personnel who have been promoted. They're employed by the laser manufacturer (the vendor of laser equipment) to work with potential clients before the sale is made—defining the client's actual needs and working out the fit with the laser manufacturer's products. Needed, Seaman says, is a combination of technical knowledge and sales orientation, along with extremely good listening skills and the ability to make presentations.

Generally these persons hold a four-year degree, possibly from a technical college, in electronics, mechanical design, or a similar discipline. Speaking and writing abilities are vital, Seaman says, because the applications engineer must be able to hold his or her own in management arguments and must, in fact, help to convince the client company that the vendor's product is the right one to buy. Even though the applications engineer is not always officially part of the vendor's sales team, he or she plays an important role in making the sale. Applications engineers must be able to identify the customer's problems and come up with workable solutions that need the vendor's products to make them feasible.

Salaries for applications engineers are in the $20,000 range for those just beginning their career at this level. Experienced applications engineers can receive up to $30,000 or $40,000 in salary. In general, they are not on commission.

Slightly higher than the salary level of an applications engineer, but probably working in a corporate research laboratory rather than for a vendor of laser equipment, is a laser scientist. Laser scientists have a laser background. A typical job here might involve "marrying" a laser to a robot. While the applications engineer employed by the laser vendor might receive from $20,000 to $40,000 a year, laser scientists can earn from $30,000 to $60,000. A few laser scientists become part of organizations that evaluate acquisitions—working for people who buy and sell laser companies.

The salesperson, or field representative, is probably a design engineer. Employed by the laser equipment vendor, he or she almost certainly holds an engineering degree from a four-year school. Added to that is a strong business orientation, including knowledge of business law, marketing techniques, and business administration. Although the salesperson does not have to have come up the ladder of experience as a service person and applications engineer, many do. Such a person has picked up the business law and marketing expertise—perhaps in night school courses—after the technical background and experience.

In one of the bigger laser companies, a salesperson or field representative could start at $30,000 and work up to $60,000 or $80,000 a year as district sales manager.

Holding a position comparable to that of the district sales manager (and earning approximately the same compensation) is a product design engineer, sometimes called an advance design engineer. Such a person has a university degree, Seaman says, and may have completed graduate work in engineering. He or she may work for a vendor, or such a person may be a scientist/entrepreneur who owns a small laser company.

JOB DESCRIPTIONS

The levels of involvement listed above may have different names in different companies. Whatever the job is called, you can be sure that somewhere there is a job description giving basic desired qualifications.

John Ruselowski, corporate sales director for Raycon Corporation, has provided four such sets of qualifications used by his company. By reading them, you'll have an idea of what persons holding those jobs actually are required to do and know.

Laser Product Technician: A person who could satisfy the required qualifications for this position would:

1. Be able to read circuit diagrams and blueprints;
2. Have a general mechanical/electrical aptitude;
3. Have or be able to develop a knowledge of the primary process for which the machine was developed and be able to optimize that process;
4. Be working in the assembly area, preparing the machine for delivery and run-off.

Laser Applications Technician: A person who could satisfy the required qualifications for this position would:

1. Become very familiar with a machine's capabilities and be able to use those capabilities in a variety of applications;
2. Have a working knowledge of computer numerically controlled (CNC) equipment, and machine tools;
3. Have a complete knowledge of CNC programming and be able to exercise the machine to its capability;
4. Have metrological and metalography knowledge for determination of laser sample dimensional and process quality;
5. Have a PC (personal computer) working knowledge for data organizing and result posting;
6. Be working in the applications lab.

Laser Field Technician/Engineer. A person who could satisfy the requirements for this position would:

1. Read circuit diagrams and blueprints;
2. Troubleshoot industrial electrical and electronic equipment on a module/board level;

3. Have a general mechanical/electrical aptitude;
4. Have a working knowledge of and be able to perform basic troubleshooting on a variety of machine CNCs;
5. Have a PC working knowledge for posting such items as repair requirements, hours, and parts replaced;
6. Be working in the field at the customer's site.

Qualifications for each of the three jobs described above require a knowledge of laser safety. In addition, persons holding those jobs must have a knowledge of industrial lasers and be able to remove and clean optical components, align the laser optical resonator for peak output, adjust focusing optics, operate the laser controller, and change flashlamps or discharge tubes for both solid-state and gas lasers.

Laser/Electro-Optical Product Engineer. Under the supervision of the Product Manager-Lasers, this person is responsible for the engineering concerned with the YAG and CO_2 lasers and associated laser equipment that is incorporated in the company's machine tool product lines. This means involvement in the quotation stage to recommend a laser product that would satisfy the customer's needs, as well as recommending and overseeing mechanical and electrical design details that are associated with the incorporation of the laser into the machine tool product.

Other associated duties include documentation of design effort. The person holding this position must generate design specifications, test specifications, and test reports. In addition, the laser/electro-optical product engineer must generate quality control procedures, safety procedures, and service procedures.

Other duties in this position include assembling and generating mechanical and electrical drawings, a parts list, and user and maintenance manuals for the laser and associated laser equipment. Other people may help with the drawings and manuals if the machine tool's operation overlaps that of the laser.

To qualify as a laser/electro-optical product engineer, the person holding the job should have a bachelor of science degree in electrical engineering, with an emphasis on electro-optics. In addition, four years of experience with laser equipment is required. One year's experience should be with the integration of lasers and machine tools. This product engineer will have had laser training at various manufacturers' locations and, preferably, schooling in computer numerical control (CNC).

A LASER MACHINIST

In Phoenix, thirty-six-year-old Rick Jackson has had over nine years' experience with lasers at Allied-Signal Aerospace, Garrett Turbine Engine Company Division. The story of how he was chosen for his position in the company's development department illustrates the idea of working your way up to running a laser.

As a Grove City, Ohio, teenager, Rick spent his last two high school years at Paul C. Hayes Technical School there. After graduation, the school helped him get a job in a Columbus factory.

Rick was offered the option of three training programs. Gear grinding and boring mills paid more money. "But I was looking for the technology of the future," he remembers. "Numerical control (NC) machines were just starting up. I thought I could get in on the ground floor." Rick trained on NC drilling and boring machines and, he says, got a pretty good background.

But Rick was young and not ready to settle down. He left the factory and spent several years working his way around the country, taking whatever jobs he could find. By the time he was married and expecting his first child, he settled in Phoenix, hiring on at what was then Garrett Engine Turbines.

"After a year, I worked my way up from manual drills to NC machines," Rick says. "My technical school background and previous work experience helped me get the promotion.

"Three years later, I heard rumors we were going to put in a laser machining center. The idea of learning new technology had always been challenging and had served me well. Long before the laser actually arrived, I kept asking to be assigned to it. When it came in, I got the chance."

That was 1979. Today, however, the company has four laser machining centers. Two are on production lines. However, in the company's development department, where Rick works, he uses an

Nd:YAG pulsed laser for cutting applications. In the company's research department, he says, workers are testing a high-powered CO_2 laser for heat-related processing.

Rick generally works six days a week, from 7:00 A.M. to 3:00 or 3:30 P.M. When he arrives, he first starts up the console. While the machine is loading software, Rick gets a cup of coffee. By the time he returns, the machine's power supply has turned itself on and is warming up the circuits.

Next Rick references the laser. He needs to align the laser beam at the exact reference spot from which all cutting dimensions are determined.

A nozzle assembly on the machine holds the focusing optics that control the laser beam. That assembly rides up and down on a worm screw. The laser nozzle has no rotating parts.

In order to help position the machine, Rick has hooked up a TV camera that runs along the same path as the laser beam when it is machining. The TV screen is mounted right on the same assembly as the console. On the screen, Rick can see electronic cross hairs that help him place the beam in just the right spot.

To reference the machine, Rick positions it over a piece of scrap metal. He can place the machine precisely—within about 1/10,000th of an inch of where he wants it to be. Then he fires a single "shot" (or pulse of energy) with the laser to mark the metal part. The TV screen shows where the laser beam "hit." Rick needs to align the site hole to the electronic cross hairs on the TV screen. Once he has matched them precisely, he has a reference point.

All instructions to the machine as to just where the laser beam will cut are based on the distances from that reference point. These instructions have been worked out beforehand and programmed into NC tape. The tape goes through a tape reader, and the instructions are stored in the CNC microprocessor that runs the laser. Because the instructions can be retrieved and loaded into the computer's memory, Rick can modify the program if necessary, just as a secretary might edit a letter whose text has been stored in a word-processing program.

Usually the company's programming department writes the software programs that tell Rick's laser exactly where and how to cut the metal engine parts. That's because his machine uses two and three axes simultaneously as it makes the cuts in the metal engine parts. Rick, however, has enough computer knowledge to modify the program or to write one himself if the programming department is backed up.

Though much of Rick's computer-programming knowledge has been picked up on the job, he says he's learned a lot from a recent CNC programming course run by Arizona State University and taught to company machinists at their request. "Our company is very education-oriented," he says. "I'm taking college courses now to become more familiar with specialized materials and processing techniques."

Because Rick works in the company's development department, he handles a number of innovative designs from different divisions. "We prove out whether the work can be done," he says. "We get parts from not only one department, but from all over the plant." Since the company makes engine parts for airplanes such as Lear jets, Rick may test the practicality of laser cutting on everything from research combustors with thousands of holes, to sheet-metal casings for the engine, with perhaps 30 holes, to gears that need clearance holes drilled into them.

Though Rick's supervisor assigns priorities and plans Rick's approximate schedule, no two days are ever the same. Sometimes he spends 90 percent of his time running the laser, making the cuts as desired. Other days he encounters problems—and must take time to solve them, slowing down his output. Each metal Rick works with, from titanium to the exotic alloys used in aerospace manufacturing (Inconel, Hastelloy, Waspalloy, and Astrolloy), has its own characteristics, which may vary according to the grade of metal or alloy used. Rick must adjust his laser to achieve optimum cutting performance on each of them.

Though he could leave the laser running during his half-hour lunch break, he usually takes it out of cycle. Safety precautions he observes include shielding for the engine parts, since molten material is ejected from the laser cut. Rick wears laser safety glasses for the specific wavelength of his Nd:YAG pulsed laser. They have been ground to his prescription; a change, since until recently he had to wear laser safety goggles over his prescription lenses. The special glasses protect his eyes from any stray laser light or reflection. Like all company workers, he wears safety shoes on the job, a standard requirement in manufacturing.

The salary for a job like his, Rick says, is anywhere from "the low 20s to over $30,000 a year—based on the length of service and a person's qualifications." Laser operators on a production line who just load parts and run machines are usually paid $8 or $9 an hour, he says, but salaries can run to $40,000 or $50,000 a year for people who work in research facilities.

Opportunities for jobs like Rick's or laser jobs on the production line are available for women and minority workers, Rick says. "Our company is real good about it. There's absolutely no reason why a qualified woman shouldn't work with lasers. In fact, I trained the most recent one myself."

The diversity of working with laser development, rather than on the production line, pleases Rick, and he hopes to continue. "Every day is a new challenge," he says. "That makes it exciting!"

LIMITATIONS OF LASERS

At Whirlpool's Benton Harbor , Michigan, Research and Engineering Center, senior manufacturing research engineer Patrick Doolan works with advanced manufacturing engineering and advanced product development and research. Only one-fourth of his time is spent on lasers, he says. A former faculty instructor in welding at the University of Wisconsin-Madison, he joined Whirlpool in 1974. He was familiar with the ruby laser in the university's mechanical engineering department and suggested to Whirlpool in 1976 that it could be used in process development work.

At Whirlpool's research and development facility, Doolan says, the Nd:YAG solid-state and CO_2 gas lasers are used. At the plants, low-powered helium-neon (He-Ne) lasers read bar codes and are part of scanning systems used to check the quality of prepainted steel. Whirlpool uses two CO_2 lasers at different plant locations for welding parts for making automatic washing machines. One laser welds a gear assembly.

The other welds a stamped bracket to the tube that spins the wash machine's basket.

"No one has lasers only as their job," Doolan points out. "Once a laser system is in place in a plant, the laser becomes a part of several people's responsibilities. In one location, a process or manufacturing engineer might have it, with a person under him doing secondary machining operations with the laser. In another location, the welding engineer might take responsibility for a laser.

"In reality, the laser becomes only a small portion of an engineer's job. They don't get to know lasers and laser technology very well. From the manufacturing and maintenance side, an operator and a lead man are actually running the laser. They are trained in basic maintenance and in how to operate the laser."

At the plant the laser operator may check tools and specifically check parts, Doolan says, but the actual loading and unloading of parts to the laser workstation is automatic. The operator watches the overall machinery—one piece of which is the laser. Since the laser is simple to run, Doolan says, an operator can bring the system up quickly. Usually the laser requires little attention. Periodic maintenance involves cleaning and replacing the optical components, and for the CO_2 laser, changing the bottles of gas.

Doolan points out that lasers represent a major capital investment for a manufacturer. Depending on the complexity of the equipment, the cost is at least $100,000, and may go as high as a quarter of a million dollars.

Generally, when lasers are chosen for a manufacturing application, they're put in to solve a special problem. "You really can't cost-effectively install a laser to replace a conventional operation that works well," Doolan says. "However, you may get a savings in cost of material. For instance, our gear assembly replaced a part made on a screw machine from a solid bar. We chose the CO_2 laser because it is the only way to make that particular gear assembly in two pieces and put it together. In our welding application, laser welding has a low heat input. Using the laser eliminated distortion and gave us quality improvement."

HOW MANY LASER MANUFACTURING JOBS ARE THERE?

Although the use of lasers in manufacturing is growing, there still are a limited number of jobs, laser veteran Seaman points out. A 1986 paper he presented at SPOT, an SME-sponsored major laser conference, estimated there were 6,000 material working lasers being used in the United States; Seaman's estimate in 1988 was 8,000 to 9,000 lasers being used worldwide in manufacturing. By the year 2000, however, that use should increase to 25,000 lasers in the United States alone— a figure that's based upon a detailed analysis of just how those lasers are being used and of future growth trends for each of the industries served by major laser industrial applications.

According to Simon L. Engle, president of HDE Systems, however, laser user companies tend to assign existing employees with many years of seniority at the company into laser processing projects. Just how many openings for new laser workers or recent graduates there will be, then, is uncertain.

For every six or seven laser machines, Seaman says, there is probably one maintenance person.

There are still fewer service personnel, since they work for laser vendors. Typically there are only two or three service personnel per company and only four or five applications engineers per major vendor.

Vendors may have only two or three jobs above that level: typically a chief scientist, a chief engineer, and a district sales manager.

Does this mean you have no chance of getting a manufacturing job involving lasers?

Not at all.

It is important to remember that these figures are estimates. As new technology develops, as more and more factories automate, as the price of lasers drops, making them more cost-justifiable to smaller laser job shops, the number of jobs available will almost certainly increase. One of those jobs may be "the job" for you.

At the McDonald Observatory, three times daily, laser beams pass out through the tele-scope, hit a reflector on the moon, and bounce back. (McDonald Observatory photo)

CHAPTER 6

LASERS IN MILITARY AND SPACE APPLICATIONS

One of the first large-scale users of lasers was the military. As early as 1965 the United States Army was developing a laser range finder. Because laser light can be used to measure distances extremely accurately, airplanes or tanks are often equipped with laser range finders. Usually the Nd:YAG laser is used to send out pulses of light to the target. When these pulses, which are invisible to the eye, hit the target, they are reflected and return to the sender. A computer carried by the airplane or tank can measure the time it took the pulses of light to make the round trip and can then calculate the distance to the target. Other range finders can be carried by individual soldiers, as they were in Vietnam by forward observers who gave distance readings to long-range artillery gunners. More likely today, however, they are incorporated into weapons.

During the Vietnam War, Air Force pilots used laser-guided bombs to destroy enemy targets, a technique Israeli pilots also used during the Middle East conflict of 1973. These "smart" bombs and rockets illuminate a target with invisible light from an infrared laser, which marks the targets. Then a laser-seeking missile or bomb can be launched and steered to the designated target.

THE STRATEGIC DEFENSE INITIATIVE (SDI)

Obviously, highly concentrated laser light can be used as an extremely destructive weapon. Since 1983 the United States has been conducting research on lasers as part of the Strategic Defense Initiative (SDI). This program, nicknamed "Star Wars," was described by former Secretary of Defense Caspar Weinberger as "designed to examine the promise of effective defenses against ballistic missiles based on new and emerging technologies. If such defenses prove feasible, they would provide for a more stable and secure method of preventing war in the future, through the increasing contribution of non-nuclear defenses which threaten no one."

The effectiveness of SDI and the enormous amounts of money being spent on it are extremely controversial issues. In an April 1987 debate between General James Abrahamson, director of the Strategic Defense Initiative Organization, and noted astronomer Dr. Carl Sagan, director of the Laboratory for Planetary Studies at Cornell University, Sagan called SDI "an immensely dangerous and foolish scheme." According to Sagan, 70 to 90 percent of the members of the National Academy of Sciences in the relevant disciplines of mathematics, physics, and engineering say that SDI can't work.

Abrahamson, however, describes SDI differently. "President Reagan did not simply propose building lasers to knock out all enemy missiles," he said. "Reagan outlined a broad program to change the entire structure of our relationship with the Soviet Union that was designed to change their behavior."

Regardless of their feelings about the ethics and practicality of the Strategic Defense Initiative, Sagan and Abrahamson agree that Soviet scientists and American scientists are studying lasers intensively. In October 1987 General John L. Piotrowski, the general in charge of American military efforts in space, told reporters that ground-based Soviet lasers already are powerful enough to destroy low-orbiting United States satellites and damage those farther away.

Today, he said, lasers at Sary-Shagan Missile Test Center in the central Soviet Union could damage a satellite up to about 750 miles high. Such lasers also could harm sensors or solar panels on satellites in geosynchronous orbit—that is, satellites that stay over one spot on earth at an altitude of 22,300 miles.

He predicted that in the future Soviet lasers could also threaten

American satellites that are part of the ballistic missile defense being developed by the Strategic Defense Initiative Office.

An October 1985 report, "Soviet Strategic Defense Programs," released by the U.S. Department of Defense and the Department of State describes the USSR's laser program as involving over 10,000 scientists and engineers, scattered over more than six major research and development facilities and test ranges.

The Soviets are conducting research in three types of gas lasers considered promising for weapons applications: (1) the gas-dynamic laser, (2) the electric discharge laser, and (3) the chemical laser. The Soviets are also aware of the military potential of visible and very short wavelength lasers. They are investigating excimer, free-electron, and X-ray lasers, the report continues, and have been developing argon-ion lasers for over a decade.

Most laser weapons need considerable power, the ability to store energy, and certain auxiliary components. Many of these have already been developed by the Soviets. For instance, they have a rocket-driven magnetohydrodynamic generator that produces more than 15 megawatts of electrical power. In addition, laser weapons need sophisticated optical systems in order to track and attack their targets. In 1978 the Soviets produced a 1.2-meter segmented mirror for an astrophysical telescope, calling it a "prototype" for a 25-meter mirror. Such a large mirror is considered necessary for a space-based laser weapon.

If technological developments prove successful, says the report, the Soviets may deploy operational space-based antisatellite lasers in the 1990s and might be able to deploy space-based laser systems for defense against ballistic missiles after the year 2000.

The basic assumption underlying SDI is that nuclear war can be prevented if the United States and the Soviet Union both maintain the ability to retaliate against nuclear attack and if the United States can impose on the Soviet Union costs that are clearly out of balance with any potential gains.

DIRECTED ENERGY WEAPONS

Researchers with the Department of Defense, working under several government agencies, are exploring technologies they feel are key to an effective defense against ballistic missiles. These technologies include directed energy weapons, designed to deliver almost instantaneously disruptive energy to the target over thousands of kilometers away. Because lasers can be retargeted rapidly and laser beams can be delivered rapidly, they could be used as interceptors to destroy the ballistic missiles and warheads before they reach space.

The United States is studying two alternatives: a space-based laser and a ground-based laser that uses space-based relays and fighting mirrors. Lasers required to "kill" incoming missiles and warheads as they are being boosted into space, as well as after they have reached space, however, require extremely higher levels of performance.

The concept of a space-based laser includes self-contained laser battle stations, according to a report prepared by the SDI Organization and submitted to Congress in April 1987. Each battle station would consist of an assembly of laser devices put together in modules. Once these stations were placed into orbit, they could engage ballistic missiles launched from anywhere on the earth, including those launched from submarines and intermediate-range ballistic missiles.

Plans for direct energy weapons technology projects call for the space-based lasers to destroy or identify decoys while they are still in midcourse flight and to defend U.S. satellites. Since the beam of some types of space-based lasers can penetrate the atmosphere down to the tops of clouds, space-based lasers may be able to help defend the United States against missiles sent from aircraft and from tactical ballistic missiles.

Since 1970 Department of Defense researchers have been studying chemical lasers fueled with hydrogen fluoride for possible use as space-based lasers. Such lasers operate in the infrared section of the electromagnetic spectrum at wavelengths of 2.7 micrometers.

Other "candidates" for space-based lasers are devices that generate beams at short wavelengths of about a micrometer or less. Because brightness increases in a ratio of 1/wavelength squared, being able to use shorter wavelengths can make the laser light much brighter if the quality of the optics and the accuracy in aiming the laser are also increased proportionally. Two of these lasers with shorter wavelengths are the radio-frequency-linear accelerator (linac) free electron laser, and the short-wavelength chemical laser. In another approach, researchers are using nuclear reactors to pump a short-wavelength laser.

GROUND-BASED LASERS

Military scientists also are exploring the concept of ground-based lasers as part of SDI. In this concept, several ground sites are equipped with laser-beam generators; with systems for locating, tracking, and pointing at the target; and with advanced subsystems for laser beam control.

These ground stations can generate a short-wavelength laser beam, adjust the beam to compensate for distortions caused by the earth's atmosphere, and project the laser beam onto space relay mirrors. Tentative plans call for placing the mirrors at geostationary orbit, so they stay in one spot 22,300 miles above the earth. The relay mirrors redirect the beams from the ground to mission mirrors at lower orbits. The mission mirrors will acquire and track the target, point the laser beam, focus the beam, and hold it on the target until enough energy is deposited to kill the target. If the concept works, ground stations located in the United States can engage targets all over the world. Such a weapon system has potential—not only for defense against ballistic missiles but also for defense against aircraft and satellites.

Also being tested are ground-based laser systems that can track and hit missiles. One such system uses the Mid Infrared Advanced Chemical Laser, developed for the U.S. Navy by TRW. A beam director, built for the Navy by Hughes Aircraft Company, provides the capability to track the target and point the laser beam at the missile.

In September 1987, a high-energy laser system at the White Sands Missile Range in New Mexico successfully destroyed a drone target missile in flight. The missile was flying at a low altitude some distance away from the laser installation. However, the laser beam destroyed vital components of the missile, which then went into a spin and exploded when it hit the ground. This test, and others, are part of a project to assess the potential of a high-energy laser system that could be used to defend ships.

JOBS IN MILITARY PROGRAMS

Working with lasers as a member of the U.S. Armed Forces is another way in which men and women can find careers in laser technology. The first step in joining the Armed Forces is to talk with a recruiter. Your high school or career center counselor can put you in touch. Or you can check your local telephone book for listings. You can sign up at 17 with parental consent; 18, without it. More than 92 percent of young men and women being recruited have a high school diploma. In addition, you'll need a personal record good enough to stand a thorough background check.

You'll be given the Armed Force Qualifications Test, as well as the Armed Services Vocational Aptitude Battery. Your scores on those tests, as well as your interests, will determine what career paths are open to you. So will the particular needs of the service to which you are applying.

When you see a career counselor at the military entrance processing station, he or she will check your scores and show you a computer screen. "On the first screen you see will be the categories the service really needs to recruit people for—at the highest level for which you qualify," says Lt. Col. Greg Rixon, Army spokesman. "The computer is updated daily to reflect current needs."

If you are joining the Army, you might see a screen about Category 39E—a job classification with duties involving intermediate-level maintenance and repair on special electronic devices, such as mine detectors, battlefield illumination systems, warning systems, and various sensors. Although you will undoubtedly work with equipment other than lasers, laser technology will probably be involved in some of the sensing mechanisms. Training required for categories such as 63E (M1 tank systems mechanic), 45Z (armament and fire control maintenance supervisor), or 41B (topographic instrument repair specialist) also will probably include working with laser technology.

"We're giving recruits a contract we will honor," says Army spokesman Rixon. "For instance, if you sign up to be in Category 39E, you are going to be one—as long as you complete the training we offer for that category."

MEASUREMENTS IN SPACE

Using lasers to measure distances in space has helped scientists learn more about the universe. Complicated mathematical models of how the universe acts have been devised, and extremely precise laser measurements help calculate "real" ranges. By comparing the predicted distances with those that have actually been measured, the earth's orientation can be monitored. Because the number of permanent lunar laser ranging stations around the globe is increasing, scientists can look for plate tectonic motions that will aid in predicting earthquakes. Data from these ranging stations are also helping to determine tidal effects and changes in the earth's rotation (shown by a change in the length of a day).

In the 1950s a group of physicists at Princeton University suggested using powerful, pulsed searchlights on the earth to illuminate mirrors placed on satellites in orbit. By photographing a satellite's position with respect to the fixed background of stars, they hoped to analyze the characteristics of the satellite's motion while in orbit.

Once lasers had been invented, however, the scientists' plans changed. Because laser light has a precise wavelength, and because laser light could be produced in incredibly short pulses, measurements in space could be made with remarkable precision and accuracy.

LASERS ON THE MOON

In 1969, as part of their moon mission, Apollo 11 astronauts placed special reflectors on the moon's surface. Later, scientists from McDonald

Observatory at the University of Texas at Austin aimed pulses of light from a ruby laser at these reflectors. The length of time it took the laser pulses to hit the reflectors and return was measured extremely carefully. Then scientists calculated the distance to the moon (238,857 miles between the center of the earth and the center of the moon.) Because the electronic circuits that measured the time were so sensitive, scientists believe the distance to the moon has been accurately measured to within two or three meters—approximately less than 100 feet. New permanent lunar laser ranging stations have been added: at Grasse, France; at Mount Haleakala on Maui; at a site 30 miles south of Canberra, Australia; and at a site near Wettzell, in southeastern West Germany. Mobile lunar laser ranging systems are operating in the Netherlands and in West Germany.

JOBS WITH LASERS AND SPACE

Finding employment as a laser technician or scientist in the space program or related fields may not be simple, since funding cutbacks are curtailing hiring. However, if working in this field interests you, you can contact university researchers, such as those at the University of Texas at Austin, or at other institutions studying these problems. You can certainly ask to be placed on the mailing list for bulletins, publications, and news releases about developments. And you may be able to use the information you get from reading the releases to contact some of the scientists mentioned directly. Sometimes, an individual scientist is willing to correspond with a young person, offering advice and suggestions. For specific job openings, write to personnel offices of programs, outlining your qualifications and asking to be advised of hiring procedures.

CHAPTER 7

LASERS IN COMMUNICATIONS

One of the most exciting developments in lasers has been their use in communications—a use that is likely to increase over the next five to seven years. "The optics boom is just starting to explode," says the director of systems technology at Xerox. "Optics in the 21st century will be what electronics represents in the 20th century."

Although the optics industry earned perhaps $10 billion in 1986, it is expected to grow at a pace of 30 percent to 50 percent annually for the next several years—fast outstripping the anticipated gain of just 5 percent to 10 percent for electronics products. Light, traveling as photons, or tiny packets of radiant energy, can move quickly. And this light can carry enormous numbers of digital signals at high speed over long distances.

LASERS AND FIBER OPTICS

Where do lasers fit into this new technology?

Laser light is a key part of the new technology. In light-wave systems, information can be transmitted as pulses of highly focused light from tiny lasers—carried over hair-thin fibers of transparent glass or plastic. At the other end of the light path, photodetectors (tiny devices) convert the light pulses to electrical impulses that can be processed by conventional techniques.

AT&T BELL LABORATORIES

Bell Labs, the immense corporate research facility that survived the 1984 court-mandated breakup of AT&T, has been a player in lasers since the earliest days. It was 1960 when Ali Javan and coworkers at the Bell Telephone Laboratories first operated the helium-neon gas discharge laser—some months after Ted Maiman had made the first working laser at Hughes Research Laboratories. Bell Labs scientist Arthur Schawlow (who later became a cowinner of the Nobel Prize in 1981 for his work in laser spectroscopy) and his brother-in-law, Charles Townes (a Nobel winner in 1964 for his work with the ammonia maser), were awarded a significant patent in 1960—a patent that they subsequently licensed to laser manufacturers. By 1961 Bell Labs had developed the continuous-wave solid-state laser (neodymium-doped calcium tungstate). And a significant advance in light transmission of information came in 1970, when AT&T scientists developed tiny solid-state lasers capable of emitting usable amounts of concentrated light continuously at room temperature. By 1977 AT&T had installed the world's first light-wave system to carry voice, video, and data communications traffic in Chicago.

LASERS AND UNDERSEAS COMMUNICATIONS

In fall 1988 the first trans-Pacific lightwave cable was placed into use, sending high-speed information under water. Additional systems for the western Pacific Ocean and the Caribbean Sea were scheduled to go into operation in 1989.

Hawaii 4/Transpacific 3, costing $600 million, includes over 5,000 nautical miles of optical fiber cable, installed by AT&T Communications between California and Hawaii, and then to a branching unit, 2,820 nautical miles from Hawaii. KDD (Japan's Kokusai Denshin Denwa Company), the Japanese international telecommunications company, has installed the branching unit, as well as 832 nautical miles of fiber cable to Guam and 1,245 miles to Japan.

Each transoceanic system uses two pairs of single-mode light-wave fibers. They're driven by a 1.3-micrometer laser that operates at 296 million bits per second. The two pairs of fibers connect Hawaii with the branching unit. From that unit, one pair of fibers goes to Guam and the other to Japan. A separate fiber-optic pair connects Japan and Guam. By 1991 the ninth transatlantic underseas telecommunications cable is scheduled to be in operation—a joint project of AT&T and four international communications companies: Teleglobe Canada, British Telecommunications International, the French Secretariat of State for Posts and Telecommunications, and Telefonica, the national telecommunications company of Spain. The major portion of TAT-9 is designed to operate at 565 megabits of information per second over each of two fiber pairs—twice the capacity of the Hawaii 4/ Transpacific 3 cable.

ADVANTAGES OF LIGHT-WAVE SYSTEMS

Light-wave systems are already carrying 24,000 telephone conversations per pair of fibers and are able to transmit data at 2.7 gigabits per second. That's fast enough to transmit the entire *Encyclopedia Britannica* in two seconds.

Light-wave communications have several significant advantages over earlier, conventional copper transmission systems. In those, digital pulses can travel for only about a mile before they need to be regenerated by electronic terminals. In Bell Lab experiments, light pulses have traveled through more than 100 miles of glass fiber without amplification. Commercial technology now in use, however, hasn't quite reached that standard; nevertheless, light pulses are traveling along glass lightguides for several miles before regeneration is needed.

Costly new construction can be avoided by using light-guide cables, since they weigh only a fraction of conventional copper cables and are only a small fraction of the diameter of the copper. That's why they can often be installed in the same ducts as existing copper cables. Since glass fiber light guides don't conduct electricity, electrical interference, like that generated by power lines and lightning surges, has no effect on the quality of transmission.

And by using a device called a *light-wave multiplexer* that combines beams from 10 lasers, Bell Labs researchers in 1986 demonstrated a system with a record 20 billion bit-per-second capacity over a 42-mile length of fiber. The 10 lasers used have generated light signals in the 1.55 micron range, according to researcher Anders Olsson— while the multiplexer has the capacity of combining up to 22 different wavelengths of laser light.

Additional work at Bell Labs in 1987 brought them the world's record for the highest transmission rate for coherent light-wave systems—sending 2 billion bits per second for 1,760 kilometers without a repeater. "Now we have to create more powerful lasers, better electronics, and more sensitive detectors to transmit at higher bit rates without significant penalties," says Alan Gnauck, a member of technical staff in the Lightwave Systems Research Department at Bell Labs.

Unlike direct detection systems, coherent systems (which use laser light) "step-down" incoming optical signals from optical to microwave frequencies. A special receiver adds light from a local oscillator laser to the incoming optical signal and produces an intermediate frequency from these two light streams. The intermediate frequency is sufficiently low so that signals can be processed efficiently by conventional electronic components. Coherent techniques and conventional electronics improve receiver sensitivity as well, because sharp electronic filters can be used.

Bell Lab researchers are also studying the feasibility of using laser light to help develop more cost-effective satellite transmission systems. If earth station antennas could be located far from a user's terminal equipment, there would be less chance of having interference from microwave communications. "We haven't used light-wave systems before," says Bell researcher John Bowers, "because we couldn't turn lasers on and off fast enough to handle a satellite's mid-to-high frequency transmission rates. But a new generation of high-speed lasers and photodetectors now allows us to do that." The system Bowers and others are studying also lets them simultaneously send satellite signals in a mix of different formats and rates over a single optical link, instead of having to split the signal up into segments and send it through parallel cables.

FUTURE RESEARCH

Bell researchers continue to work with lasers—in particular, tiny semiconductor lasers that drive photons through the optic fibers. These devices, made of materials with electrical properties between those of conductors and insulators, produce the intense, very pure beam of light when stimulated. Because of their small size, high speed, and minimal power consumption, they're practical for light-wave systems. With AT&T Technologies semiconductor and laser plants in Allentown and Reading, Pennsylvania, and with a fiber-optics plant in Atlanta, Bell has a firm commitment to the technology.

JOBS AT BELL LABS

Technical employment at Bell Labs is under the direction of Lloyd Friend. Competition for positions is high; *Fortune* magazine reports the staff includes 3,430 Ph.D.s, and 80 percent of the students offered jobs at Bell Labs accept. The labs have been a leader in recruiting talented women and minorities as scientists and engineers. They sponsor a number of scholarships, fellowships, and grants and offer summer employment as well. For further information on such programs, write to Bell Laboratories, 150 Kennedy Parkway, Short Hills, NJ 07078.

LASERS AND OPTICAL STORAGE

You may already know that lasers play an important role in compact disc technology. Since their introduction in 1983, the audio compact disc player and CDs have grown into a worldwide market with more than $1 billion in annual revenues. The compact disc has billions of microscopic pits on its aluminum surface. In these pits, music is stored in digital form. During playback, a laser beam scans the pits as the CD is spinning inside the player, sending the information from the pits to a computer chip, which converts it into sound.

Optical data storage, however, offers more advantages than merely high-fidelity music reproduction. An optical disc that's smaller than the familiar computer floppy can store the equivalent of a quarter of a million pages of typed information.

Originally, optical discs had ROM (read-only memory). Now, companies are developing erasable-disc technology. In May 1988 Tandy Corporation announced plans to license rights to the dye-polymer technology, which uses a high-intensity laser beam to encode a disc that's specially coated and a lower-power laser beam to read back the information. Hitachi and Sony are working on similar technology. Sony's disks, introduced in late 1988, were competing with those from Verbatim (a Kodak subsidiary), Sharp, and Pentax. R&D experts in other corporations are working on erasable disks for computers that are based on a technology that uses a laser for changing magnetic properties. At stake: a market estimated at $8 billion by 1991.

Even today's optical disc technology is speeding computer operations. Banks using optical discs from Bell & Howell find that information from multiple workstations is quickly accessible.

For instance, when a customer comes in to apply for a loan, the person taking the information can tap into the bank's mainframe database and use its information to help assess the creditworthiness of the applicant. Credit histories, credit checks, and other documents relating to that customer may be scattered throughout bank files. Optical disc storage and retrieval, however, lets the loan officer access the information randomly and quickly.

What does this technology mean to you?

If you want to work in this fast-paced, competitive environment, you must keep up with developments. Read magazines like *Time, Business Week, Fortune,* and *Forbes.* Check library indexes for the *Wall Street Journal.* Find the major companies researching this technology, write for annual reports, get on the company mailing list for press releases, and write to the appropriate personnel offices to learn of job openings and required qualifications. Because of intense competition among the companies, skills in sales and marketing will be important tools as you look for jobs—along with the ability to understand the technology and to help predict its advantages. As *Time* magazine reports, "The alluring glow of optics is pointed straight toward profit and increased productivity."

HEWLETT-PACKARD LASERJET PRINTERS

The company Dave Packard and Bill Hewlett started in 1939 in a garage behind the Packards' home in Palo Alto, California—with $583 in capital—has grown to one of the top 100 industrial corporations in America, with net revenue of over $8.1 billion. Although HP makes more than 10,000 products, one of its best-known is its desktop laser printer. The LaserJet, LaserJet PLUS, and LaserJet Series II printers work with more than 600 of the most popular software programs. Introduced in 1984, when the company "broke the barrier of the $100,000 laser printer" by inventing the desktop LaserJet for under $5,000, the printers have been a runaway best seller. In fact, more people own LaserJet printers than all other laser printers combined, giving HP approximately an 80 percent market share.

"A laser printer is based on copier technology," explains Jeri Peterson, press coordinator, Hewlett-Packard Boise Printer Operation. "There's an internal laser in the printer. When you type information into your PC via a software package, that data is transferred to the printer. It controls where the laser beam writes by exposing an area on a round photosensitive drum.

"As the laser beam moves across the drum, it exposes just a tiny dot on the drum. That laser has the capability of defining 300 dots across and 30 dots down in each inch, giving it quality almost equal to that of a daisywheel printer.

"As the laser exposes the part of the photosensitive drum, it changes the electrical charge on the drum. The laser moves quickly while the drum rotates into another area within a toner cartridge. The toner, which has an opposite charge, is attracted to the area where the laser wrote on the drum.

"As the drum rotates, it passes over the paper which is underneath the drum. Beneath the paper is a corona wire, carrying a charge opposite to that of the toner. Consequently, the toner is attracted to the paper. Next, it travels through a fusing unit (395°F) that melts the toner to the paper."

One big advantage of laser printers is that they're quiet, flexible, and fast, since you can print up to eight pages per minute. Because they're not tied to a font ball, like a daisywheel printer, users have the ability to integrate text and graphics—in effect, doing desktop publishing.

A PRESS RELATIONS COORDINATOR

Jeri Peterson, 36, didn't expect to be handling press relations for HP's Boise Printer Operation when she finished two years at a junior college in Yakima, Washington. In fact, she wasn't sure just what she wanted to do, so she tried several fields. She was a nanny for a California family, a hospital phlebotomist in a medical lab, an *au pair* in Paris where she worked for her room and board while taking French classes, and—eventually—a college student again when she realized she needed a degree. Earning her bachelor's degree in fine arts from the University of Washington, Seattle, she signed up for a campus recruitment interview. HP was looking for an industrial designer, but hired Jeri as a graphic designer in the disk memory division that makes large disk drives for HP's minicomputer systems. By 1984, when HP introduced the new LaserJet technology to dealers, a new system of marketing was needed; Jeri was on the ground floor. HP uses a "Dealer Channel" group as support for dealers. "I supported the Eastern region, doing dealer presentations," Jeri remembers. "I wrote proposals and documentation for desktop publishing applications."

By March 1987, when HP introduced the LaserJet Series II (a second-generation printer), the company essentially was making two earlier models obsolete. New sales promotion, effective for dealers, was needed if HP was to meet its goal of adding unit sales to make up for rolling over the higher-priced models. A time crunch came up, and Jeri found herself coordinating 14 separate promotion projects.

Today, based in Boise, Idaho, Jeri deals mainly with the computer trade press. She's responsible for talking with editors when they phone, finding out what they need, and setting up interviews with on-site management people. Surveys show that PC-oriented consumer publications and trade press reviews play an important role in the user's buying decision, so the impact of successful public relations is strong.

With HP planning a key product introduction in 1989, Jeri's

responsibilities also include planning time. "We want the chance to change from *reactive* public relations to *pro-active pr*," she says. "But that doesn't happen overnight."

Since HP has flexible hours, Jeri picks her own starting time, generally choosing to begin work at 6:00 A.M. If editors are visiting, she dresses more formally; otherwise, a casual dress is fine, she says. One recent day found her awaiting an editor who wanted to write about HP's technical service phone line and who wanted to interview the college students who work part-time and take as many as 20,000 phone calls a month from HP customers. Before the editor arrived, Jeri had briefed the service line's manager on kinds of questions the editor might ask and how to handle sensitive issues. After a 90-minute meeting with the manager, she made sure all was ready for the continental breakfast she'd planned.

Jeri sat in on the interviews—not to censor them, because she generally doesn't even talk during them, but to make sure that the HP manager covered key issues. "If there's confusion, I'll restate the question," Jeri says. All went smoothly, but she also sat in as the editor spent an hour with a student who was manning the lines. By the time she returned to her desk, she found 10 pink slips—phone calls from editors of other publications and from market analysts. "Although I'm not the HP spokesperson on strategic issues," Jeri says, "I do spend time talking with my managers, getting them to return calls. I brief them on who the editor is, so that when an article does appear, the HP message generally comes across."

Another component of Jeri's job is responsibility for product introduction public relations. HP currently uses a Los Angeles-based agency for news release writing; Jeri coordinates with the agency, often traveling there to discuss sales promotion and press ideas.

With 40 percent of her time spent on phone calls, and 20 percent in travel, Jeri has a long workday. Although she tries to leave by 6:00 P.M., she has a computer and LaserJet printer at home. She often works weekends at home but generally won't come into the office. Her short-term goal: to free up time to do more product introduction planning.

Salary for a job like hers, she says, ranges from $20,000 to $40,000 but would be higher if she were with an outside public relations agency.

"That's because agencies good with high-tech marketing need qualified people," she explains. Although she does not have an MBA, she recommends that degree for someone who wants the business side of laser products. Her advice to young people starting out? "Don't be afraid to try things, even if they're outside your area of expertise. Stretch and risk a little in what you are doing."

Sales, marketing, public relations, dealer relations, promotion, and publicity. All these are nontechnical areas, Jeri says, but they're ideas in which people wanting work related to lasers can find enjoyment and careers.

HOLOGRAMS

Holograms have been described as painting with light. That's not strictly true, of course, but to people who watch Dr. Tung H. Jeong, Albert Blake Dick Professor of physics at Lake Forest (Illinois) College and author of *Laser Holography: Experiments You Can Do*, it seems as if that's what he's doing. Magically, through using a simple laser and everyday objects, Jeong demonstrates that laser light can make a color picture on black and white film without a camera—a picture that can be seen with ordinary light.

An inexpensive videotape by Jeong showing the technique as he explains it to young students is available from the Thomas Alva Edison Foundation, 21000 West Ten-Mile Road, Southfield, MI 48075. So is the book he wrote for students.

The process looks simple. Yet, Jeong says, four separate Nobel Prizes have been awarded for the theories contained in that one sentence.

In order to make the process easier to understand, Jeong uses what he calls the "soap bubble theory." He shows kids how, with water just from the sink, it's possible to produce beautiful soap bubbles with all the colors of the rainbow. It's the same theory, he says, that lies behind how we get holograms in color from black and white film.

How Lasers Make Holograms

When Jeong makes a hologram, he splits the laser beam in two with a lens. One of the two beams bounces off the object Jeong is using and is reflected back onto photographic film. The other beam from the laser hits the film directly, but doesn't bounce off the subject. Although both halves of the beam were coherent when they left the laser, they are no longer coherent when they reach the film. Some of the light waves that strike the film arrive in certain patterns, producing a double-strength wave called a *reinforcement.* Other waves show a pattern producing a *cancellation.* These patterns are recorded on the film. The resulting hologram can be used to reconstruct a three-dimensional picture of the object.

One of the most fascinating things about holograms is that you can cut up one of them into tiny pieces. Yet, unlike ordinary photographs, each small fragment contains a complete representation of the object.

How Holograms Are Used

You may be carrying a hologram right now. Just look at a credit card. If you see a multicolored three-dimensional image of the MasterCard or Visa logo, you have a transmission hologram. Actually, the embossed insignia really does not transmit light from the far side of the hologram, but the silvery backing fools the hologram into thinking it's doing so. Another common use of holograms, though you may not realize it, often happens at the supermarket check-out counter. The clerk passes one of your items over the scanner window. A spinning hologram under the counter locates the bar code on the product. The hologram directs the reflected light back from the bar code into the store's computer, which has been programmed to recognize the item and to enter its price on the cash register.

Corporations also use holograms as an attention-getter. Jeong himself was commissioned to photograph former Olympic gymnast Mary Lou Retton for possible use of the hologram by McDonald's, one of her sponsors. "We went to Salt Lake City for the project," Jeong recalls, "since I had a friend there with some special equipment." Retton was photographed with a light exposure of 1/20 billionth of a second, with a special lens under vibration-free conditions. Jeong created a similar photograph of Ronald McDonald, used in McDonald's annual franchise show at Las Vegas.

Holograms have many other uses. One that might not occur to you can be thought of as similar to time-lapse photography. As Jeong reported at a 1988 meeting of the International Society of Optical Engineers, once you've made a hologram, you can compare what you recorded with the living, and growing, object. For instance, he says, you can superimpose a hologram image of a mushroom on top of the growing fungus. You can measure the growth, second by second, as it grows in real-time—as you watch the differences between the static hologram and the living fungus.

This ability to compare recorded versus actual objects makes using holograms helpful in nondestructive testing. An aircraft tire can be recorded on a hologram and then inflated. If there is a defect in the tire, it will expand at a slightly larger rate than the rest of the tire. The trouble spot shows up, when compared to the baseline hologram.

Holograms can even show antimatter. Working with Jeong, scientists at Fermilab near Batavia, Illinois, have recorded bubbles formed in a liquefied hydrogen chamber—particles that exist for only 10^{-12} seconds. The bubbles that form for such a brief time can be photographed by laser light.

Information storage is another technology Jeong and other researchers are pioneering. One of his friends from China is able to record entire encyclopedias on a single sheet of film—pages that can be randomly accessed. With such a technology, it would be possible to store your entire personal medical record on a piece of plastic the size of a credit card.

Another application for holograms has been to record cultural treasures. Treasures from the Soviet Union and from the ancient civilization of Thrace have been photographed with laser light, and their resulting holograms are being sold as art objects. It was Soviet scientists, in fact, that worked out the process by which holograms are made that are visible under "normal" white light; a second kind of hologram, invented earlier, could only be viewed by laser light.

HANDS-ON WORKSHOPS

At Lake Forest College, near Chicago, an annual five-day Holography Workshop each July teaches participants the techniques for making a wide variety of holograms. No previous experience or scientific background is necessary, says Jeong, who directs the workshop. Information is available from Lake Forest College, Lake Forest, IL 60045.

In addition, the college hosts an International Symposium and Exhibition on Display Holography every third year. In 1988 the symposium featured scientists and artists from 18 nations who presented the latest technological and artistic developments in holography. A juried show featured 50 works by many of the world's top holographic artists.

A HOLOGRAPHY ARTIST

One of the jury members for that show was Doris Vila, chairman of the Department of Holography at the School of the Art Institute and herself an accomplished holographer. One of her latest commissioned holographic works is at the School of Nursing, University of Wisconsin-Eau Claire.

Vila, who says she "came at" holography from the artistic, rather than the scientific, side, became interested in the subject even before she saw her first hologram in 1979. "I wanted to know more," she remembers. Now, she views holography as an art form, rather than a revolution in imaging technology.

She explains holograms to students by asking them to imagine that they are "standing at the edge of a pond, throwing in a stone that generates wavelets out in circles. When you throw in a second stone, it too has circles of wavelets. At a certain point where wave crests generated from each stone meet, where crest meets crest, it gets higher, and where crest and trough meet, they cancel each other out. Imagine that we can make a small metal grating that captures the interference pattern. We'd insert it into the pond, and let everything go still again. Yet the grating would record those patterns."

As Vila explains, you can think of holograms as a tiny window (or the small grating in the pond). The information as to the stone's position is contained in a field over the holograph. It's not a specific point-to-point correspondence, as it would be in a photograph. That's why holograms are essentially a storage medium.

At the School of the Art Institute, in Chicago, courses are offered in beginning and in intermediate/advanced holography, with a smaller number of students going on to independent study. Even the beginning course is a full semester of hands-on learning, in which students set up their own cameras, tune spatial filters, and work with their own personal imagery. Although the emphasis is on artistic results, students do study the structure of light, as well as the theory and techniques of three-dimensional imaging. If they wish, they can sign up for as many as 24 hours of lab time each week.

Advanced students gain a working knowledge of multi-image display holograms, as well as techniques for producing master holograms, both for white light transmission and reflection work.

Students in the MFA program can emphasize holography.

"The School of the Art Institute has the most substantial art holography program in the United States," Vila says. Other resources she recommends include New York Holographic Labs and the Museum of Holography in New York City, as well as Chicago's Museum of Holography and a similar museum in Los Angeles: Holographic Visions.

Vila herself specializes in large-scale rainbow holography. "I work in a narrative style, combining photographs, shadowgrams, and stenciled imagery with found objects," she says. "What artists find quickly is that you need to learn the scientific principles behind holography in order to do better work. What photography taught us about how we see, holography can teach us about how we perceive."

CHAPTER 8

LASERS IN RESEARCH

Another area of laser technology in which jobs exist is scientific or industrial research. Working in laboratories, scientists explore possible new uses for lasers and ways of improving current applications.

TECHNOLOGY TRANSFER

Technology transfer—to and from military and space programs—is taking place in a number of locations, including the various military service research and development agencies, NASA centers, and federal laboratories. Here, significant advances in technology and recent inventions from the military programs are being studied. Scientists hope they will be able to use them commercially. Many of these items are already being produced by the private sector. Others will become new products of tomorrow.

Technology generated by the Strategic Defense Initiative (SDI) is bringing a broad range of spin-offs. These can add up to significant benefits in terms of human welfare, industrial efficiency, and economic value.

Qualified American business and academic clients that have been approved under procedures established by the Department of Defense can learn about civil applications of this technology through using a referral database. Open to all federal and state agencies, the database is accessible through a computer modem. Technology application panels are being established in various areas, including biomedical applications;

electronics, communications, and computer applications; power-generation, storage, and transmission applications; and materials and industrial process applications.

MEDICAL FREE-ELECTRON LASER PROGRAM

Since 1985 Congress has funded medical, biomedical, and materials research on free-electron laser technology as part of the SDI budget. Regional medical free-electron laser research centers have been established at Stanford University, the University of California at Santa Barbara, Brookhaven National Laboratory, the National Bureau of Standards (in Maryland), and Vanderbilt University.

In addition, preclinical medical research on surgical applications, therapy, and the diagnosis of disease is being conducted at the Massachusetts General Hospital, the University of Utah, Northwestern University, Baylor Medical School, and the University of California at Irvine. Biophysics research is being carried out at the University of Michigan, Purdue University, the University of Texas, Jackson Laboratories (Maine), and Physical Science (Massachusetts). Materials science is being investigated at Brown University, the State University of New York at Buffalo, the University of Utah, and Stanford, Vanderbilt, Princeton, and Southern Methodist universities.

OTHER SPIN-OFF APPLICATIONS

Key examples of SDI technology that have potential civil applications provide a substantial economic return on investment. They include optical computing, using laser light instead of electrical circuits for transmitting data; more efficient, less expensive electrical power systems; lightweight mirrors that can be aligned through computer control and used for lasers in manufacturing processes; and integration of laser technology, robotics, and computerized techniques for precision control into applications for manufacturing processes and biomedical work.

In addition, free-electron lasers have the potential for being used in noninvasive cancer surgery, early diagnosis and treatment of heart disease and stroke, and other medical diagnostic and treatment applications.

Scientists in the laboratories mentioned above are working hard at finding new and practical uses for lasers. You can learn more about what they are doing by writing to the public relations or press office of the universities and hospitals, asking to be put on the mailing list for copies of news releases mentioning lasers. Addresses for the universities can be found in reference books available at your school or public library and in inexpensive almanacs.

LASER FUSION

Research into laser fusion offers another career opportunity for working with lasers. If laser fusion energy can be commercially feasible, it could provide an environmentally safe form of energy that is virtually inexhaustible.

The sun generates its energy through thermonuclear fusion of hydrogen atoms. Fusion energy research on earth is an effort to recreate and harness that energy. In the sun gravity holds charged particles together in a tightly packed mass. That's why fusion reactions can occur on the sun at temperatures of about 14 million degrees. But because the earth has only a fraction of the sun's enormous gravity, scientists must create more extreme conditions in order to make fusion possible.

On the earth the fuel density of a fusion reaction must be in the range of 10 to 20 times that of lead, and temperatures must reach about 50 million degrees. When these conditions have been achieved, the fusion fuel undergoes a thermonuclear "burn." As a result, large amounts of energy are released—many times more energy than the laser beam emitted to start the reaction.

At the University of Rochester, in New York, the Laboratory for Laser Energetics uses a 12-trillion watt OMEGA laser system to study the potential of high-powered lasers to produce controlled thermonuclear fusion. In order to do this, target pellets of fusion fuel must be heated and compressed so rapidly that the fusion fuel will burn before the highly compressed hot material flies apart. Powerful laser beams, split, amplified, and converted from infrared light to ultraviolet (which is more effectively absorbed by the target pellets) are focused precisely on the pellets. When the beams hit the pellets, the surface matter blasts outward at a velocity of nearly 600 miles per second. An equal force implodes on the shell containing the fuel. The kinetic energy of the imploding material is converted to heat.

The OMEGA laser is the size of a football field. The beams it emits arrive at the target within a millionth of a millionth of a second of one another at a spot defined by dimensions smaller than a tenth of the diameter of a human hair.

In March 1988 scientists at Rochester reported they'd achieved a major milestone by using the OMEGA laser to compress and heat a small capsule of fusion fuel to the highest density achieved that has ever been directly measured—in the range of two to four times that of lead, with a temperature in the range of 5 to 10 million degrees. The fusion fuel was compressed to a density more than 100 times its normal liquid density. If you could compress water to the same degree, an eight-ounce glass would weigh about 50 pounds. A gallon of water, compressed to the same degree as the fusion fuel in the laser experiments, would weigh nearly half a ton.

Sophisticated technology allows the scientists to split each of OMEGA's 24 laser beams into several thousand beams that strike the target simultaneously. The target pellets are glass shells about the size of a grain of sand, containing a frozen layer of fusion fuel. In order to compress the pellets evenly and prevent an area on their surface from "ballooning out," the laser must irradiate the entire surface of the spherical fuel pellets with a high degree of uniformity. Splitting the beams so they strike the fuel pellets simultaneously lets the laser do this.

The burst of energy from the OMEGA laser that compresses and heats fusion fuel is incredibly short. The blast of laser light lasts about .6 of a nanosecond (6/10th of a billionth of a second). For that period of time the OMEGA laser is 20 times brighter than the peak generating capacity of all the electrical generating plants in the United States.

Scientists from the U.S. National Laboratories reviewed and endorsed the results of the University of Rochester's laser fusion target experiments. Design and engineering studies on upgrading the OMEGA laser have been partially funded by Congress. The upgrade, estimated to cost $39 million (in fiscal 1988 dollars), would raise OMEGA's energy from 2,000 to 30,000 joules—and, scientists hope, would enable them to achieve the high densities and temperatures it takes to ignite fusion fuel.

Additional research in laser fusion is being conducted at Lawrence Livermore National Laboratory, Livermore, California. Scientists there are using the extremely large Nova laser for similar inertial confinement experiments.

LASER ISOTOPE SEPARATION

Scientists at the Lawrence Livermore National Laboratory are also using lasers to separate isotopes in order to increase the concentration of valuable forms of the elements uranium and plutonium. Their goal is to help keep the price of American enriched uranium competitive with the price charged by foreign companies.

In its natural state uranium is a mixture of two isotopes: U-235 and U-238. But U-235 amounts to only about 0.7 percent of natural uranium by weight. Uranium used as fuel in a nuclear reactor must have three percent of U-235. Consequently, natural uranium must be enriched.

Because different uranium isotopes absorb light tuned to different wavelengths, laser light—precisely tuned to desired wavelengths—separates the isotopes. The system uses two types of lasers: dye lasers that generate the light used for photoionization of the uranium and copper-vapor lasers that energize the dye lasers. Powerful green-yellow light from the copper-vapor lasers is converted to red-orange light in the dye laser. This red-orange light is tuned to the precise colors that are absorbed by U-235 but not by U-238.

In the Livermore project, uranium is heated in a vacuum chamber. A set of laser beams, tuned to wavelengths that match those of the desired U-235 atoms, passes through the vapor. When the atoms absorb the laser light, they pick up enough energy to give up one of their negatively charged electrons. The U-235 atoms—now positively charged—are pulled from the vapor by an electric field and become enriched uranium, which can be made into fuel to drive nuclear power reactors. The U-238 atoms, which don't have an electrical charge, pass through the electric field and onto a collector.

CONTRACT RESEARCH

Not all research with lasers is taking place at universities. Typical, perhaps, of the type of arrangement possible between industry and a research partner is the contract between the Gas Research Institute, an industry trade association, and SRI, a consulting group from Palo Alto, California. Current technology used in detecting gas leaks is based upon a flame ionization device. The laser-based leak detector SRI is developing, however, may reduce labor costs for leak surveys by as much as 50 percent, as well as improve pipeline safety.

"We put out a competitive Request for Proposals (RFP)," says Dr. Tom Altpeter, research manager, environment and safety, at Gas Research Institute. "We were testing the technical marketplace of ideas to see if we could find something better than existing technology."

What GRI wanted was a detector that would be selective for ethane (natural gas), a detector that would ignore methane produced by swamp gas, or gas from automobile exhausts.

As Altpeter explains, the device, when tested and operational, will find natural gas leaks from buried pipes in the street or gas mains. The laser will be mounted on a van, which can be driven at speeds up to 30 miles per hour. As the laser beam sweeps the street, at a distance up to 150 feet ahead of the van, its beams fan out, becoming divergent, rather than being compressed into the more common, threadlike laser beams. The carbon dioxide laser, which normally operates around the 10 micrometer range, has been converted to the

3 micrometer range, which is selective for ethane and ignores other gases.

Light from the laser is reflected back to special receiving devices on the van. If the laser light senses ethane, however, it's preferentially absorbed by the leaking gas. Careful measurements, built in as part of the device, show the difference between the light the device has sent out and the light reflected back when ethane is present.

Testing of a prototype device is under way. If the laser detection device proves successful, it will hit commercial markets about the early 1990s.

LASER RESEARCH AT BATTELLE

At Battelle Memorial Institute, more than 7,500 scientists, engineers, economists, and supporting specialists conduct more than 2,000 studies per year that cost about $400 million. Much of this work is done by Battelle under contract from industrial organizations and government agencies.

Laser research at Battelle-Columbus is handled by a special laser technology group. In an assortment of programs, Battelle is developing applications of high-powered CO_2 laser radiation for welding, cutting, transformation hardening, and cladding of metals. In addition, lasers are being studied for improvements in cutting, shaping, and thermal processing of glasses and ceramics.

"At Battelle, we're working with low-powered and high-powered lasers," says Dr. Frank Jacoby, principal research scientist. A "sensor group" is researching applications for solid-state and He:Ne lasers. At Battelle, other scientists run a CO_2 laser lab in which industrial research is carried on.

Battelle scientists are also studying two types of Nd-doped lasers: an Nd:YAG laser, which is used for testing various industrial processes, and an Nd: glass laser. "One of our laser scientists believes that by using modern glass and making the laser beam very long and thin that we can achieve the same thermal performance as we do with the Nd:YAG laser," explains Jacoby.

The large glass laser Battelle owns also is used for simulation in a government-funded "Star Wars" project. "The beam is focused down to simulate what might happen if someone ever built a Star Wars big laser," Jacoby says. "We're studying the materials interaction."

Still other laser projects at Battelle include laser shock hardening. When a laser beam is shot into metal, it can make the metal 60 times stronger, Jacoby says. That idea has gone beyond the testing stage; one Battelle scientist has received a patent, and metal companies are working with Battelle to commercialize the process.

JOBS IN RESEARCH

Research scientists who work with lasers have challenging careers. On the one hand, they have the knowledge which puts them on the frontier of technology; on the other hand, funding for R&D projects is often dependent on outside sources, such as Congress or private companies. Consequently, how many jobs there will be and who will get (and keep) them may depend on how successful the laboratory is at coming up with proposals or landing contracts.

One way to learn about such jobs is by reading the trade publications. For instance, laser researchers were recently needed by Lincoln Laboratory, a federally funded research center operated by the Massachusetts Institute of Technology. An ad running in *Lasers & Optronics* described research efforts at Lincoln Laboratory as involving the development of electro-optical systems for laboratory and field experiments, systems that used state-of-the-art lasers, computers, sensors, and electronics. Applications included high-powered laser beam control and target detection and discrimination for the national defense.

Researchers at Lincoln Laboratory were developing multidimensional laser radar systems in order to classify, discriminate, and track both tactical and strategic targets. These laser systems use image processing, binary grating optics, and high-speed data processing.

In order to qualify for these, and other positions, candidates should have a doctor's or a master's degree in physics, electrical engineering, or mechanical engineering. In addition, they should be experienced in such fields as quantum optics, atmospheric physics, optical engineering, image and signal processing, or circuit design.

SECURITY CLEARANCE

Because many research laboratories are working on government-funded projects, U.S. citizenship is generally a requirement for employment. In addition, your personal record should generally meet any requirements for obtaining security clearance. Laboratories like Lincoln are equal opportunity employers and welcome applications from qualified candidates, regardless of sex or minority background. Salaries are competitive, and benefits are generally comprehensive.

In labs like Battelle, entry-level jobs for laser technicians require a two-year program in technology, according to Jacoby. Typically, the technician runs the equipment. The next level up, for Battelle, is that of laser researcher—a man or woman who often runs experiments as well as doing day-to-day laser operations. Most researchers Battelle hires have a four-year degree; however, Battelle recently promoted one laser technician who attended Ohio State University (with tuition help from Battelle) and who received the bachelor's degree.

Next highest position? Project researcher, if you're managing projects. Generally, the master's degree is the prerequisite. Persons like 46-year-old Jacoby, who have a Ph.D., can become a research scientist, who is basically in charge of medium-sized projects, or handles a certain subsection of a major project.

Jacoby says that unlike company labs in which defense contracting plays a major role, Battelle doesn't require U.S. citizenship and often has a number of foreign nationals working. "That's because we have so many projects happening simultaneously," he explains.

A LASER RESEARCH SCIENTIST

One of the roles Jacoby finds himself playing is that of fund-raiser. "At Battelle, principal research scientists like me are essentially people who get ideas, think up projects, and then go out and try to find people to pay for them," he explains. "For instance, I have the CO_2 laser lab. In our laser paint-stripping research, we run tests, and then we try to find people who want to strip paint."

Jacoby's days usually start at 8:30 A.M. and end at 5:30 P.M., though in research labs, he says, the hours are somewhat flexible. On the days he meets visitors and nonlab personnel, he dresses in a suit and tie; "If I'm down messing with the equipment," he says, "I come in wearing grungies. What I wear depends on my schedule."

A typical recent day found Jacoby helping to host a visiting group from a tractor company interested in laser robotics. Though the group spent the entire day at Battelle, he spoke with them for about two hours, leading them on a tour of his part of the facilities. Two or three more hours that day were spent in proposal writing and in catching up with his correspondence. Later, he helped troubleshoot in another lab, helping the scientists there to solve problems on their project and arranging for equipment to be built for their future needs.

Although Battelle has no hard-and-fast rule, Jacoby says, roughly 40 percent of his time "should" be spent in marketing Battelle's services. That might involve travel to prospective companies that might fund projects or arrange for Battelle to do contract research.

Salaries for scientists at his level, he says, range between $40,000 and $60,000—higher on the coasts, because of the cost-of-living difference. Jacoby says that senior research scientists in laboratories might earn as high as $70,000.

His advice to young people who want to work with laser research: look for universities or hospitals with laboratories and get hired at the technician level. Ohio State University has a research foundation similar to Battelle's, he says, in which companies with a problem can hire a team headed by an OSU professor to work on it. Such a lab affiliated with a university is a good source of openings for laser technicians, Jacoby says. "It's a good way to start in the laser area."

CHAPTER 9

EDUCATION AND TRAINING

What skills do you need if you want to work in laser technology? What training should you have, and how should you get it?

Because laser technology careers encompass many fields, there is no right answer to those questions. However, there are several factors you will want to consider.

START EARLY

Junior high school is not too early to be thinking about math and science competency. Because much work with lasers involves your understanding of concepts in physics and optics, you will want to prepare yourself to take courses in those subjects.

In 1983 a U.S. National Commission studying education proposed basic standards for high school graduation that included four years of English and three years each of social studies, mathematics, and science. In December 1987 William Bennett, U.S. Secretary of Education, had even more specific proposals. Bennett suggested high school graduation requirements that included three years of math chosen from the following courses: algebra I, plane and solid geometry, algebra and trigonometry, one semester of statistics and probability, one semester of precalculus math, and calculus. In science, Bennett suggested three years of courses, chosen from astronomy, geology, biology, chemistry, physics, or principles of technology.

How practical is such a requirement? What can you do if your high school does not offer this wide range of courses? In a big city school system such as Chicago or Detroit, not every high school has courses in physics or calculus, since the "pool" of interested students academically qualified to enroll may not be large enough in a particular school to make it feasible. Usually, however, these courses are offered in "magnet" schools or at least are available in a technical high school within the large city system.

If you think you want to work with lasers, then, you will have to take an aggressive role in your education. You will have to work closely with guidance counselors, math and science teachers, and other school personnel to be sure you take advantage of every opportunity. Perhaps there are special exams you must take to qualify for a magnet school or a technical high school. A teacher or counselor may be willing to give you extra coaching or tutoring so you can do well on the exams; may plan extra assignments or suggested projects you can complete outside of class; or may be sure you are aware of enrichment opportunities. Such a person can put you in touch with a faculty member from a nearby college or university or perhaps can arrange a visit for you with someone working in lasers.

You may have to make the first move in letting your teachers and counselor know you are interested. Once they realize from your attitude and willingness to work hard in school that you are serious about a possible career in laser technology, they will be glad to help. Even if they don't have answers to all your questions, they can direct you to experts.

In high school, then, take all the math and science you possibly can. You will want to do well in English, also, since communication skills will be helpful in making reports and presentations. Since computers are playing an increasingly important role in today's world, you will want to become as proficient as possible in computer skills.

Your guidance counselor is a good resource for you. He or she can help you learn about colleges and universities offering courses in optics and engineering and can work with you to plan your high school curriculum to meet entrance requirements for specific schools.

One educational seminar for high school students offered by SPIE— The International Society for Optical Engineering, in cooperation with the Thomas Alva Edison Foundation is "Frontiers of Optical Engineering." In 1987 this seminar was run concurrently with OE/LASE, an important trade show. At the seminar SPIE members talked about optical engineering, as well as the academic preparation required for a career in this field. Although topics vary from year to year, some subjects presented at past seminars have included the making of a hologram, lasers and how they work, high-speed photography, optics for astronomy, time-reversing mirrors, and computer-aided tomography. For more information on these seminars, write to SPIE, PO Box 10, Bellingham, WA 98227-0010. The telephone number is (206) 676-3290.

ADDITIONAL PREPARATION

If you didn't take enough math or science in high school, don't be discouraged. You can still catch up. You will need to work closely with counselors and advisers to find out just what your deficiencies are and to plan a program of study that will help you become proficient. If you're planning to take physics or chemistry, and you didn't have those courses in high school, you will want to talk to teachers and counselors well in advance, to be sure your math skills are adequate. If they're not, you may want to take the appropriate math courses before enrolling in those sciences. Today, many colleges have specific placement testing for math courses that can pinpoint your strengths and weaknesses in specific skills. Tutoring, math labs, and study helps such as workbooks, audiocassettes, and videotapes can aid you in achieving your goals.

COURSES IN OPTICS

Each year, SPIE surveys colleges and universities in the United States and Canada that offer two-year and four-year programs in optics and related fields. Their findings are published in a special report: *Optics in Education.* In 1986-87, 50 schools offering these programs were listed.

Optics includes far more than lasers, of course. But most of the schools included in the SPIE report do have coursework in laser technology. You can get an idea of the importance of lasers at a particular institution by reading the listing for that college.

A typical listing for a college will include facts about how the optical program is organized—that is, whether courses are offered by a special department, or whether they are taught in various college divisions. Not all schools are alike. For instance, at the Center for Applied Optics, University of Alabama in Huntsville (which offers bachelor's degree programs in optical sciences and in optical engineering), all departments of the School of Science and the School of Engineering have optics courses and a strong optics orientation. Within the School of Science, the Departments of Physics, Mathematics, Computer Science, and Chemistry are coupled to the optics program; while in the School of Engineering, the participating departments are Mechanical, Industrial, and Electrical and Computer Engineering. At the University of Virginia, Charlottesville, Virginia, however, course work and optics are offered by the Department of Electrical Engineering.

Knowing this information can be useful. You can request catalogs and admissions information from the appropriate college personnel and learn early just what the specific requirements are for admission in the program of your choice. Then you can plan your high school course of studies to meet those requirements.

Listings in the SPIE *Optics in Education Report* also describe facilities. For instance, Western Washington University, Bellingham, Washington, has a laser laboratory that students can use for independent study in addition to regular classwork. In the past, students have built their own lasers, which are now part of the laboratory. At Austin College, Sherman, Texas, undergraduate students majoring in physics have laboratory facilities available that include helium-neon, nitrogen, and dye lasers.

Optics in Education also includes information about college/industry relationships. Arizona State University, Tempe, Arizona, is close to Phoenix—which has companies involved in lasers and laser materials processing. Students may find opportunities for part-time employment or may find it easy to meet and talk with people already employed in laser technology. At Camden County College, Blackwood, New Jersey, the laser program uses a committee of local industry experts to help plan its curriculum and to help obtain donated equipment. Past laser graduates presently working in industry also have helped the program.

Reading the listings in *Optics in Education* will also give you descriptions of many of the courses offered or at least a summary of what's available in different schools. At Oregon Graduate Center, Beaverton, Oregon, research is in progress on semiconductor lasers for a variety of applications in communications, remote sensing, and integrated optoelectronics. At New Mexico State University, Las Cruces, New Mexico, you can take a course in "Laser Physics" from the Physics Department or a course in "Introduction to Lasers" from the Electrical and Computer Engineering Department.

From *Optics in Education,* you can learn numbers of students in various programs at the different schools, where opportunities exist to study lasers at the undergraduate or graduate levels, and information about the faculty presenting the courses.

A current copy of *Optics in Education* can be obtained from SPIE at PO Box 10, Bellingham, WA 98227-0010. Additions or updates are included in a column feature in each issue of *OE Reports:* "Optics in Education."

COLLEGE COURSES TO TAKE

"If you want to do graduate study in optics and lasers, any good university with a good program in physics or electrical engineering can prepare you," says Dr. M. J. Soileau, who heads CREOL—the University of Central Florida's Center for Research in Electro-Optics and Lasers. "You really need a solid background in either physics or applied physics. You need a very strong math background, starting with calculus. As a college undergraduate, take as much math as you can, at the highest level of sophistication possible. The more you're proficient in math, the easier it is to do the science and the engineering. A good background in physics and engineering—preferably electrical engineering—will help you if you want to do laser research."

PASADENA CITY COLLEGE

Many colleges and universities that offer courses in optics and lasers have quite tough requirements for entrance. One that does not— deliberately—is Pasadena City College, Pasadena, California. From the very start of the program in 1977 the laser classes at Pasadena City College have allowed all comers into the program. The school believes that this open admissions policy is the best screening technique available.

The laser electro-optics technology program welcomes students from a wide range of backgrounds. Many come directly from high schools with technical and scientific preparation. Some are majoring in physics at the community college and want to know about lasers and optics. Other students in the program already have four-year and five-year degrees. Still others are physicians who are interested in lasers and who believe the program will show them how lasers can be used in their jobs.

Because the school has an extremely high job placement rate for graduates, and because Pasadena City College is a two-year program, the need has arisen for those laser electro-optics technicians working in Southern California companies to continue their education. The University of La Verne, a small private university about 25 miles away, has worked closely with Pasadena City College to develop a four-year program that offers a bachelor's degree in optical engineering.

Although some community college graduates need to take additional classes in calculus, engineering physics, or chemistry to prepare them adequately for upper-division work at the University of La Verne, once they have achieved required competency, they are encouraged to take the upper-division optical engineering courses. These include classes in "Advanced Laser Design and Construction," "Advanced Optical Systems Design," and "Thin Films in Optics."

UNIVERSITY OF ROCHESTER

The Laboratory for Laser Energetics, a multidisciplinary teaching and research unit of the College of Engineering and Applied Science at the University of Rochester is the first of its kind at any American college or university. Students are involved in all of the research programs, including a project to explore the potential of high-powered lasers to produce controlled thermonuclear fusion as an alternative energy source. The laboratory's principal research tool is a 12-trillion-watt laser system. At the laboratory, research activities include major programs in photo-matter interactions, optical materials development, laser physics and technology, and the physics of ultrahigh density phenomena.

The laboratory contains the new Ultrafast Science Center, which investigates the production and utilization of phenomena occurring on time scales of less than a billionth of a second.

Undergraduate students who want to concentrate in engineering are assigned faculty advisers in the College of Engineering and Applied Science in their freshman year. They may (and usually do) begin taking engineering courses as early as their first semester.

During their first two years, they receive a strong liberal arts education. In the spring of their sophomore year students apply formally to the College of Engineering and Applied Science by filing a "concentration approval" form in which they list an approved plan of study for their last two years.

A special five-year program is available for electrical engineering juniors who contemplate graduate work. Students are accepted into this program in the spring of their junior year and can begin master's level independent work in their senior year. At the end of the five-year program, both a bachelor's and master's degrees in electrical engineering are awarded.

Institute of Optics

The Institute of Optics at the University of Rochester is an internationally known center for teaching and research. Optics majors who plan to do graduate work may apply in their junior year for admission to the five-year B.S.-M.S. program.

Students normally apply for admission to the Institute of Optics at the end of the sophomore year by submitting a concentration approval form. Certain prerequisite courses and certain minimum cumulative and specific grade point averages are required.

Interested and qualified undergraduates are often able to take part in faculty research projects during the school year or in the summer. Current projects include studies involving lasers, holography, image processing and information handling, experimental studies of optical and electronic properties of matter, computer-aided lens design and evaluation, design of multilayer optical filters, and the interaction of intense optical radiation with matter, including studies of laser-induced nuclear fusion, a potential energy source for the future.

The Institute of Optics, in cooperation with the Rochester Institute of Technology and Monroe Community College, has been designated as *the New York State Center for Advanced Optical Technology.* The center is part of the state's program of centers of advanced technology, designed to foster university-industry research in areas of extraordinary potential for translation into industrial growth.

From the broad range of research and development activities under way at the Institute of Optics, six areas most likely to result in the rapid transfer of technology to industry have been chosen for the initial program of the center. They are:

- Integrated optics and fiber optics
- Optical fabrication-gradient index optics
- Opto-electronic systems for image recognition, medical optics and robotics
- Optical materials including thin films
- Phase conjugation techniques in laser development image evaluation, digital image processing, and color science

Prospective students and undergraduates considering optics as a major are encouraged to write or to visit The Institute of Optics, University of Rochester, Rochester, NY 14627.

UNIVERSITY OF ARIZONA

At the University of Arizona, Tucson, the Optical Sciences Center is a graduate center for research and teaching in optical sciences and engineering. It has numerous laboratories as well as equipment used for research in a broad range of optics.

Interdisciplinary programs currently involve the departments of Astronomy, Chemistry, Civil Engineering and Engineering Mechanics, Electrical and Computer Engineering, Microbiology, Physics, Physiology, Planetary Sciences, and Radiology.

Several faculty members have specialties in laser work—including laser system application, laser spectroscopy and laser spectroscopy of solids, holographic techniques, laser physics, pulse propagation in laser amplifiers and attenuators, short-pulse production in lasers, high-energy lasers, and laser amplifiers.

Says faculty member Dr. Frederic A. Hopf, "While second-harmonic generation of high-power pulsed lasers and low-power continuous-wave lasers is a well-developed technology, little is known about high average-power applications.

"The ultimate goal of one of our research projects is the development of high-power (more than 1 kilowatt) single-mode visible lasers.

"Our free-electron laser (FEL) project has concentrated on an investigation into the nature of short-pulse traveling-wave amplification."

At the Optical Sciences Center, work on laser analysis has included development of computer models of laser fusion, laser isotope separation, free-electron lasers, and other complex laser systems. The free-electron laser model was the first three-dimensional, time-dependent treatment and included detailed interactions between the optical and electron beams.

Dr. George N. Lawrence, associate professor, describes one OSC research project as "studying the design and performance of optical data storage read/write heads, which consist of a laser diode, a slab waveguide, and a chirped grating to couple from the waveguide mode out to a focused spot."

Also under study: theory of laser isotope separation in polyatomic molecules and ionization of atoms in very strong laser fields.

UNIVERSITY OF CENTRAL FLORIDA

One major center for graduate students is the UCF Center for Research in Electro-Optics and Lasers (CREOL)—an interdisciplinary center formed to provide direct access by Florida's high-tech industry to UCF's program in electro-optics. Faculty members from the departments of electrical engineering, physics, math, and mechanical engineering work with the program, along with a board of directors that includes several prominent representatives of local industry. Florida has one of the largest concentrations of electro-optic industrial activity in the United States, with some 15 laser companies in Orlando alone.

"We hope to establish CREOL as the third major national center for research and education in lasers," says Dr. M. J. Soileau, who heads CREOL. "Although students enter other academic departments, earning their master's degree in electrical engineering, physics, or math, they do their thesis or dissertation with one of the faculty members associated with CREOL."

Research is an integral part of CREOL. Soileau and several faculty members are studying nonlinear optics, including a project on how very high-powered laser radiation interacts with materials. Related projects include using high-intensity light to do optical switching—studies that may have eventual application to optical computing and optical information processing. "In addition," says Soileau, "our studies can lead to defense applications that help protect people from deliberate or inadvertent exposure to lasers. This work involves a variety of different kinds of lasers. With pulse lasers, short pulses act as clocks like a strobe light to stop or study the details of very fast processes in nature."

Another CREOL faculty member is using various kinds of crystals to change the color of laser light—research that may have medical implications. "You may have the type of tumor that absorbs light preferentially at one wavelength," Soileau explains. "So you want to tune the laser to that wavelength. This project is attempting to come up with new laser sources that can be tuned over a large span of wavelengths, but can be compact and economical. Although we currently can make tunable lasers, they tend to be extremely large. We need new sources for smaller tunable lasers."

Another group of CREOL faculty is studying propagation of light beams through turbulent media. Essentially, they hope to use lasers to remotely measure pollutants in smokestacks. Also interested in their research: the Department of Defense, which wants to learn how sending laser beams through various atmospheric conditions will affect results.

"Much of modern optics involves thin films of material," Soileau says, "films that are only a fraction of a micron thick. Thin film optics are used throughout any laser application. They can be antireflection coatings to reduce spurious light. They can be beam-splitters to split off certain portions of laser light and recombine it again. The problem with lasers, though, is that laser intensity is high enough that fragile films fail. At CREOL we're starting a program to develop a technique to deposit very thin films for use in lasers."

A Laser Research Project

The story of Edesly Canto, 27-year-old graduate student from Panama, illustrates what it is like to work with lasers on a research project.

Edesly brings a rich and varied background to her studies. By the time she finished six years of elementary school in Panama, she qualified through examinations and high grades for a science high school. In her six years there she studied chemistry, biology, and mathematics and took several levels of physics classes as well. By the time she graduated she was fluent in Spanish, Portuguese, and English. She enrolled in Southeastern Oklahoma University, Durant, Oklahoma, and earned a bachelor's degree in physics, with minors in chemistry and math. During her studies she did a number of optics experiments that involved small lasers. She studied dispersion properties of lasers and learned how lasers work.

Next, she became a teaching assistant at North Texas State University. She completed her master's degree in 1985 and became research assistant to Dr. Soileau, then a North Texas State Professor. When Soileau moved to Orlando to head up CREOL, Edesly came along. She had finished her academic work for a Ph.D. at North Texas State and will receive her doctorate from that university after her CREOL research project and dissertation have been completed and approved.

At CREOL Edesly studies nonlinear properties of semiconductors, using zinc selenite. "We're trying to study the decay parameters and transport phenomena for this material," she explains, "so we can determine zinc selenite's physical constants. Once we know the constants, we will know how zinc selenite will behave under radiation from the atmosphere. We will know whether it can be used for windows on jets and other aircraft."

Edesly works with lasers daily, using the laser as a tool to study zinc selenite. "The Nd:YAG laser is a picosecond pulsed laser," she says. "I work in a time scale of picoseconds—one millionth of a second. By probing the sample of zinc selenite with this kind of time scale, we can study its microscopic properties. The pulses of the laser are so short that we can model what is happening in the sample."

The laser Edesly uses looks like a black box about five feet long and two feet wide. Most of the box is hollow to protect the actual path of the laser light from being scattered around. Even though the light itself can't be seen, since it is not in the visible part of the spectrum, Edesly and other CREOL researchers must wear special goggles to protect their eyes. She puts a special kind of crystal in the path of one of the laser beams to produce a light of exactly 1.06 microns. The light coming out of the laser is approximately 2 millimeters in diameter, but Edesly can compress or enlarge the beam for different experiments.

Each morning she arrives around 8:30 or 9:00 A.M. She turns on the laser power supply, which applies voltage to the flashlamps that optically prompt the material inside the laser. "The next thing I do is check the pulse of the laser," she says. "I want it to be the right width in time—about 30 picoseconds. I want the shape to be Gaussian—a shape that looks round to your eyes, but if you scan it in space, it looks like the bell curve used in statistics. I check the energy of the laser beam to be sure it meets my standards of 4 millijoules in energy. All this must be done each day and takes about an hour to complete."

Edesly's particular experiment probes the zinc selenite material with three different laser beams, all coming from the same Nd:YAG laser. "I use special mirrors," she explains. "Just by setting the laser in different positions I can make the beam go wherever necessary. Consequently, I need to check the alignment of all my optical mirrors, making sure all beams overlap at the sample. Once they do, I observe my conjugate beam, which is the product of the overlapping of the three beams. By setting that fourth (or conjugate) beam I can get all the information necessary for my data.

"I can change the energy of each beam independently, if I like, by having an attenuator in each beam line. That's a combination of a linear polarizer and a halfway plate."

All this is computerized, she explains. Edesly has detectors that are connected to a microcomputer that makes the measurements.

She learned computer skills first as an undergraduate at Southeastern Oklahoma, but she took most of her computer courses at North Texas State University. "Although I have help from many people," she says, "we do have to modify the software we need for our work. We also need to write all the plotting." Edesly uses Pascal (a computer language) for most of her work, but when she needs to go on UCF's mainframe computer for a long program, she uses FORTRAN, another computer language.

It's extremely important, she feels, to know the safety regulations for working with lasers and to follow them exactly. If she's working with liquid crystals she wears a lab coat; otherwise she dresses in comfortable clothing. She always wears goggles when working with lasers, and if she is handling chemicals, she wears rubber gloves. Her work day depends on what she needs to get done for her particular experiment. Sometimes she spends 10 or 11 hours at the lab, and frequently she works on weekends.

When she finishes the requirements for her Ph.D., Edesly hopes to teach solid-state physics and optics at a university and to be involved in research as well. She'd rather work in lasers at a university than in industry, she says, but she believes the field is wide open. "Lasers are demanding and fascinating," she says. "There's so much more to learn. We're really just beginning to understand them. There's plenty of opportunity."

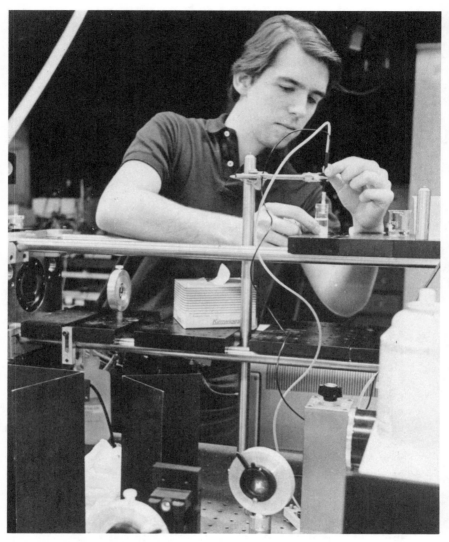

This student is using lasers for routine studies of chemical systems. (Laser Institute of Technology photo by Jerry Mitchell)

CHAPTER 10

JOB-HUNTING TIPS

How do you go about getting a job working with lasers? And what, realistically, are your chances of doing so?

They may be better than you think—if you are qualified. But you're probably going to hold another job within an organization first before you are promoted to working with lasers.

There are several reasons why this will probably happen. First, of course, a laser represents an investment on the part of the company. While it is true that technological advances have brought laser prices down to perhaps $300 for a helium-neon laser, the large industrial lasers or medical lasers cost thousands of dollars. Understandably, a beginning technician won't get turned loose to play with the company's new toy. Instead, a department head is far more likely to promote from within—giving someone already employed by the organization the first chance at working with the laser.

USEFUL STRATEGIES

Nevertheless, there are certainly strategies you can use to get yourself in a position to be considered for laser work. One of them is to be selective about the training you receive. Send for bulletins like *Optics in Education,* available annually from SPIE—the International Society for Optical Engineering, PO Box 10, Bellingham, WA 98227-0010. Study the course listings carefully. Call colleges and universities you are interested in. Ask about their placement rate for graduates, especially those who've taken coursework in laser technology. Perhaps they will give you the names of one or two alumni in your area that you can write to or phone. Persons who have studied lasers and who are now working in the field can give you invaluable advice about the preparation you should have and the current job market.

When you do enroll for coursework, choose a school or technology program accredited by the Accreditation Board for Engineering and Technology (ABET). You can get a list from the board by writing the organization at 345 East 47th Street, New York, NY 10017. The telephone number is (212) 705-7685. While graduation from an ABET-approved school or course of instruction is certainly not an automatic entry requirement for laser jobs, ABET thinks of itself as responsible for the "quality control for engineering education" and closely monitors the educational environment in the institutions it accredits.

During your school years, become active in local chapters of professional societies (see appendix A). Writing to the national headquarters of those listed will put you on the mailing list for information. Ask to be placed in contact with the appropriate local chapter. Attending meetings regularly will help you learn more about lasers and will give you a chance to talk with professionals employed in the field.

Read three or four of the periodicals listed in appendix B of this book; if possible, subscribe to them. Not only will you keep up with technological developments, but you will also learn dates of future conferences and seminars. Try to attend these. Often fees for students are extremely low—or waived, with a letter from one of your professors. At these meetings you'll find vendors of various laser products. Pick up catalogs and information literature as well as business cards of company representatives. Some of them may be willing to talk with you about laser jobs and possible employment. Or they may know of customers—laser users—who are expanding and who are hiring.

Reading the trade periodicals will also give you a list of vendors whom you can contact for literature, so you can keep up with new laser developments. Each year, several publications run a *Buyers Guide,* which includes company addresses. In addition, you can check your school or public library for reference publications listing lasers and laser services.

KEEP LEARNING

No matter whether your first job is with lasers or not, plan to continue your education. The technology is changing so rapidly that unless you devote time to learning new techniques, keeping abreast of new developments, you will not succeed.

Through the journals, as well as through the literature from the various associations, you can learn about seminars and training on lasers. For instance, the American Society for Laser Medicine and Surgery offered two short courses in Dallas: "Laser Fundamentals for Nurses," and "Advanced Concepts in Laser Therapy for Nurses." Course objectives included discussing three approaches to marketing a laser program, identifying at least two areas of potential research in laser nursing and possible methods for studying them, and interpreting new regulations into a workable safety policy.

SPIE offered 88 tutorial short courses and 16 engineering update courses at O-E/LASE '88. They included tutorials on diode lasers, excimer lasers, physics and technology, coatings for laser systems, and similar topics.

The Society of Manufacturing Engineers (SME) offers clinics about lasers, including "Lasers in Electronics Manufacturing"—a recent conference whose sessions included meetings on "Laser Spectroscopy as Probe of Material Processing Reactions," and "Microelectronic Laser Welding—Proven on the Production Floor." Companies whose key people have attended past sessions include Hughes Aircraft Company, TRW, Rockwell International, Hytek Microsystems, AT&T Bell Laboratories, Hewlett Packard, and A-B Lasers.

Attending such courses not only improves your own knowledge and enhances your chances of getting a job working with lasers, but also gives you the opportunity to talk with top professionals. Often these meetings lead to informal "networking" later on. At conferences, also, professional societies often have a placement service through which members seeking to change jobs can meet those who are hiring.

Another strategy you can use to help distinguish you from other eager candidates is to find a summer job or a co-op job in your field, as early as possible. Even if you need to be a volunteer in a laboratory, that experience will count for more when you look for your postcollege job than a higher-paying spot at a fast food outlet, suggests veteran placement consultant Howard Rudzinsky. In addition to being able to use people at the company as references, "grunt work in the lab" will give you education you may not get in any other course or classroom, he says.

USING PROFESSIONAL PLACEMENT AGENCIES

Although some of the professional journals and technical magazines have "employment opportunities" sections, using a professional employment agency is one of the ways in which companies find qualified personnel. One such agency is Louis Rudzinsky & Associates, a Massachusetts-based placement firm founded in 1968 that specializes in optics, electro-optics, lasers, and laser physics. Typical clients of the firm include Fortune 500 organizations, private research organizations affiliated with major corporations, smaller emerging companies, and major research labs.

"We're looking for professionals from the associate degree level all the way up to the Ph.D.," says Howard Rudzinsky, senior consultant. "There are several important points we look for as we review candidates."

One, of course, is how well students have done. Since graduates, especially of associate degree programs, will come in as laser technicians, optical technicians, or research associates, their scholastic achievements may be the edge that helps them land the entry-level job. Students who will graduate from an associate degree program are urged to phone or write the firm several months before commencement, Rudzinsky says. The address is Louis Rudzinsky & Associates, 394 Lowell Street, Suite 17, Lexington, MA 02173. The phone number is (617) 862-6727. Rudzinsky says he's willing to talk with graduates and "to assess the likelihood of their achieving what they desire."

"Placement isn't a one-shot deal," he says. "I'd like to be the one helping you find your job, but if not, then perhaps, your next job." Fees, incidentally, are always paid by the client corporation; there is no cost to job-seeking individuals.

Rudzinsky says being totally honest and upfront is essential when you're job-hunting. If you've been terminated, be honest about it and give the real reasons for the dismissal, he advises. Don't mislead an agency about your past salary. Don't use a placement agency merely to get an offer from another company, so you can extract a counteroffer from your own firm.

"It's a small world," he says. "Everybody knows each other. We attend the same conferences, meetings, and exhibitions. We know professors that you've studied with. Unprofessional conduct will come back to haunt you."

If you use an agency, be prepared ahead of time to answer "what if" questions about relocating. Do you insist on staying close to your present location? Will you work only on the East or West coasts? Competition there among corporations for qualified candidates is stronger, so salaries are slightly higher. You may decide to trade a salary increase for stimulating work or for the lure of an entrepreneurial culture in a just-starting-up firm.

Agencies similar to Rudzinsky's are also active in laser placement. You can read their advertisements in such publications as *OE Reports* (a SPIE publication). They include Dunhill of Framingham (Four Franklin Commons, Framingham, MA 01701, (617) 872-1133) and Lyon Associates (PO Box 568-SP, Amherst, NY 14225, (716) 836-0690).

RÉSUMÉS

When you're job-hunting, whether you're a beginner or experienced, highlight significant accomplishments. If you've achieved academic honors, list them. Instead of merely writing down the titles of courses you've had, however, indicate the types of laboratory or industrial equipment you've operated, including the types of lasers. Since many laser systems are computer controlled, put down your computer literacy—what hardware you've worked with or what computer languages you're fluent in. Many engineers work well with FORTRAN; others have experience with Pascal or "C," which is a programming language.

If you've only worked for one company, summarize your experience. Be prepared to capsulize it by listing the two or three most significant things you've done.

Don't get discouraged if you don't get hired immediately by the company of your choice—or even by any company working with lasers. "Many companies and managers keep the résumés our placement firm submits," says Rudzinsky. "They stockpile them. When R&D dollars come in or a contract shows up, they know whom they're interested in interviewing."

SALARIES

What can you expect to earn in lasers? Detailed results on engineering salaries are available (but high priced) from the American Association of Engineering Societies (AAES). Write the AAES Publications Department, 415 Second Street NE, Washington, DC 20002. Their annual surveys include data from over 200 private-sector employers, representing over 100,000 engineers. Although specific salaries in lasers are buried in the expensive reports, a four-page summary bulletin of engineering salaries is available for $12.00 plus handling/postage. Orders must be prepaid. The 1985-87 salary trends for engineers in all industries showed a median salary of $29,450 for those with no experience and a bachelor's degree. Very substantial salary differences exist among workers in the same sector and with identical amounts of experience, however. Engineers with 10 years' experience reported salaries ranging from $32,550 to over $50,200 in aerospace. Laser technicians with two-year degrees and no experience can expect to start around "the high teens to low twenties," according to Rudzinsky, depending upon their schooling and achievements.

Don't forget the "business" side of lasers as you consider opportunities in laser technology. Companies offering lasers need product sales and support staff—knowledgeable men and women who can talk to customers about their needs and can make recommendations that may ultimately result in a sale. Backgrounds useful for obtaining those positions will probably include marketing and sales training, as well as enough science and technology to understand customer problems and the products being offered.

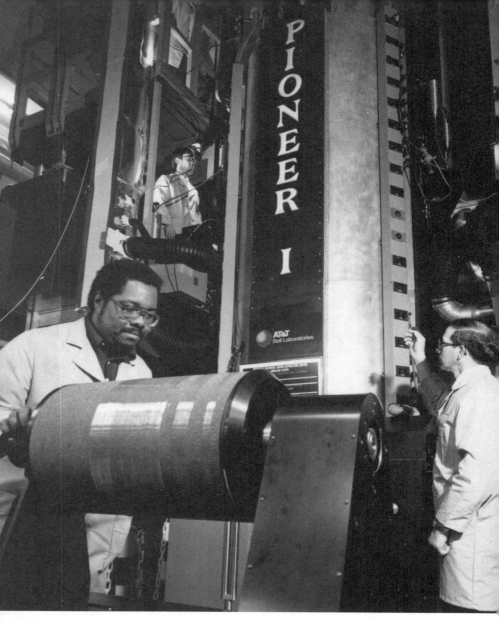

Workers test experimental optical fiber designs by creating glass preforms and heating them in a furnace. (Reproduced with permission of AT&T Archives)

CHAPTER 11

WOMEN AND MINORITIES

There's good news and bad news about opportunities for women and minorities who want to work with laser technology. The good news is that sex and ethnic origin are no barriers to qualified candidates. In other words, if you have the prerequisite skills and desire you have excellent chances of finding employment—if (and it's a big if) companies are hiring. The bad news is that proportionately fewer women, blacks, Hispanics, and foreign nationals are enrolling in engineering programs. In undergraduate enrollments the share of total enrollments by Asian Americans continues to rise.

ENROLLMENT STATISTICS

There are several reasons why women and minority enrollment in engineering programs may have peaked. One is a change in the way in which these statistics are calculated. Until 1986 the American Association of Engineering Societies (AAES), the parent organization that tracks enrollments through its Engineering Manpower Commission, made estimates to allow for missing data. Now, actual figures are used. Experts aren't sure whether there's really been a drop or whether earlier statistics were inaccurate.

Another reason for fewer women, minorities, and foreign nationals' enrollment in engineering programs is that there are, in fact, fewer college students. From 1983 to 1995 the numbers of students in that age bracket will drop by one-fourth. A nationwide survey of college freshmen indicates that although 3.6 percent of women planned to major in engineering in 1982, by 1986 that figure had declined to 2.9 percent. Nevertheless, women still account for one of about every seven engineering students.

At the graduate level, however, things are different. *Engineering Manpower Bulletin No. 86* (February 1988) on women in engineering indicates that women still earn more than 2,000 master's degrees in engineering each year. However, only a handful of women are earning doctor's degrees. Based on records from the National Research Council showing engineering doctoral awards from 1975 through 1985 to U.S. citizens whose race or ethnicity was known, only 627 women received a doctorate in that decade. Within that number Asian women earned 48 (7.7 percent), black women earned 9 (1.4 percent), Hispanic women earned 10 (1.6 percent), and white women earned 560 (89.3 percent).

Although enrollments in engineering technology programs leading to a bachelor's degree have increased for women from 4 percent (fall 1976) to 9 percent (fall 1986), women are still greatly underrepresented. At the associate degree level in engineering technology, in fall 1986 11.5 percent of all full-time undergraduates were women.

In 1985, according to the Office of Civil Rights, 1,450 women earned the bachelor's degree in engineering technology. Of these, 144 were black, 32 were Hispanic, 36 were Asian, 9 were American Indians, and 1,097 were white non-Hispanics. Nonresident, non-U.S. citizen women earned 57 degrees. No breakdown on race or ethnic background is available for the remaining 75 women who earned degrees.

Other reports confirm the proportionately low numbers of women entering the engineering profession. How many of these graduates are actually working with lasers or in fields related to laser technology is not known. Yet, more women have entered the engineering labor force from 1981 to 1986 than in all of the preceding years—an indication that for those who are qualified, opportunities exist.

ENGINEERING MANPOWER BULLETINS

These facts and others are available in annually updated bulletins that can be purchased at $12.00 apiece, plus postage and handling.

For a list of publications and prices, write to Director of Manpower Studies, American Association of Engineering Societies, Inc., 415 Second Street NE, Washington, DC 20002. The phone number is (202) 546-2237.

CONCENTRATE ON SCIENCE AND MATH

If you are a woman or a member of a minority group, what can you do to become qualified to work with lasers? One important strategy is to excel in science and mathematics. As a junior high or high school student, sign up for all the math and science possible. Often, especially in large metropolitan areas such as Chicago or Atlanta, there are opportunities for enrichment programs in those subjects. Even if such programs require your after-school, weekend, or vacation time, take advantage of everything you can. The skills you gain and the contacts you make can be invaluable. Also, as you begin to read about schools with engineering or optics programs, write to the director of admissions at those institutions that interest you. Ask about summer studies, minority recruitment, or special programs you can join.

One reason for picking up additional math and science beyond that taught in your regular high school's curriculum is that there are serious shortages of qualified high school teachers. A 1985-86 survey by the National Science Teachers Association concluded that almost one-third of all high school students are being taught science or math by teachers who are not qualified. In addition, a survey of State Title II directors showed that over 200,000 math teachers and 319,000 science teachers need further in-service training. Half the high school science teachers, the survey discovered, have had less than six hours of in-service training in science within the last year.

Statistics from the National Research Council indicate that the mathematics achievement of the top 5 percent of the 12th grade students is lower in the United States than in other industrialized nations. The average 12th grade mathematics student in Japan outperforms 95 percent of comparable U.S. 12th graders.

Math and science are two important tools you'll need to qualify for jobs working with lasers. Optics and electro-optics are demanding, "hard" sciences. Finding opportunities for additional study and taking advantage of them may give you the edge you need for admission to a top-quality college or university program.

SOCIETY OF WOMEN ENGINEERS

A major organization that tracks the achievements and statistics of women in engineering is the Society of Women Engineers. SWE offers pamphlets and information on scholarship programs, convention news, an SWE information packet, and subscription information for *U.S. Woman Engineer,* a magazine you'll find invaluable for latest findings. Send a self-addressed, stamped envelope to SWE, at its headquarters at United Engineering Center, Room 305, 345 East 47th Street, New York, NY 10017. Ask for "FACTS: An Introduction to the Society of Women Engineers." Returning the coupon found in FACTS will get you on the mailing list for additional information. Bulletins and application forms describing the 38 scholarships SWE administers are also available; send a self-addressed, stamped envelope along with your request.

Deadline dates are extremely important. Applicants for freshman and reentry scholarships can receive applications from March through June. Those who apply for sophomore, junior, and senior scholarships can get applications only from October through January.

WESTINGHOUSE CAREER LITERATURE

Special publications aimed at encouraging black and Hispanic youngsters to consider engineering as a career are available free of charge from Westinghouse. Although laser technology is not addressed specifically in Westinghouse's literature, nevertheless the material is useful for boys and girls from junior high age on up. To receive either the black or Hispanic publications (they're similar, but not identical), call Westinghouse at 1-800-245-4474 (in Pennsylvania, call 1-800-

242-2550). Listed in each are various associations along with addresses and phone numbers. These, too, can be contacted for more material. They include:

National Fund for Minority Engineering Students
Suite 3105
220 East 42nd Street
New York, NY 10017

National Hispanic Scholarship Fund
c/o Gilbert Chavez
PO Box 748
San Francisco, CA 94101

National Association of Black Engineers
905 15th Street NW
Washington, DC 20005

Mathematics, Engineering, and Science Achievement (MEBA)
Lawrence Hall of Science
University of California
Berkeley, CA 94720

SUMMING IT UP

The word from those working in lasers and laser technology is that sex or ethnic background is not a barrier to hiring. Women and minorities who want to succeed certainly can. You'll recall that in chapter 4 you read stories of a woman podiatrist and a woman physician specializing in reproductive endocrinology, as well as those of a male laser nurse/laser safety officer. Today, your opportunities in laser technology are limited only by your desires and willingness to learn.

Top: A worker at a British firm uses lasers to cut large aluminum aircraft panels in three dimensions. *Bottom:* A British company employs a computer numerically controlled laser to cut out lettering with premise repeatability. (London Pictures Service photos)

CHAPTER 12

INTERNATIONAL OPPORTUNITIES

LASER TECHNOLOGY IN CANADA

Lasers and related technology are beginning to play a more important part in Canadian employment. Research is continuing at a number of locations, including the Ontario Hydro Research Centre in Toronto. There, various applications of lasers are studied, including the use of lasers for induced chemical reactions for isotope separation. Lasers and fiber optic sensors for measuring temperatures of generators are also being studied.

Another project there that involves lasers is the use of holography to detect quality problems in tubing used in nuclear reactors.

For more information, contact Ontario Hydro Research Centre, 800 Kipling Ave., Toronto, Ontario M8Z 584 Canada, 416-231-4111.

At the University of Toronto, several noted professors are involved with research in lasers and related topics. Chemistry professor Dr. Geraldine Kenny-Wallace, Nobel Prize-winner Dr. John Polanyi, and Dr. Boris Stoicheff, a renowned physicist, are among them.

The Ontario Laser and Lightwave Centre, also in Toronto, is a research facility that is studying low-power diagnostic laser systems, especially in the sensing area. Other projects in which laser work is taking place involve nondestructive evaluation of solid state materials and methods of making high-quality optical fibers. Companies also are using facilities at the centre for basic research, supplying their own technicians and using the centre's equipment under a contractual arrangement.

For more information, contact Dr. Henry van Driel, associate director,

Ontario Laser and Lightwave Centre, University of Toronto, Toronto, Ontario M5S 1A1.

Other Canadian facilities involved with lasers include the National Optics Institute, CP 8554, Sainte-Foy, Quebec, G1V 4NS, and the Manufacturing Technology Centre, National Research Council of Canada, Montreal Rd., Ottawa, Canada K1A OF6, 613-993-2436.

Alberta Laser Centre

At the University of Alberta, research in laser technology has been pursued since the mid-1960s. Established in 1984, the Alberta Laser Institute continues research in various laser applications. It concentrates on industrial applications, but current programs involve robotics, materials processing, laser sensors, electronics, and medical applications. Facilities include state-of-the-art CAD/CAM-based laser manufacturing capabilities that include cutting, heat treating, cladding, welding, and drilling—processes that are used on materials as diverse as ceramics, plastics, rubbers, glass, wood, textiles, paper, and electronic circuit fabrication elements.

Dr. Vivian Merchant, program director for materials processing, describes the institute's work with Alberta companies as providing four components of service: process development, in which the Alberta Laser Institute studies whether or not lasers can be used cost effectively to manufacture components a firm produces; job shop work, in which companies use institute facilities for tasks such as laser welding of parts; consultants, helping companies choose laser equipment appropriate to their manufacturing needs; and research and development projects that examine new laser techniques.

For additional information, contact the Alberta Laser Institute, 9924 45th Ave., Edmonton, Alberta, Canada T63 5J1, 403-436-9750.

Additional information on technology developments is available from Alberta Technology, Research and Telecommunications, 10909 Jasper Ave., Edmonton, Alberta, Canada, T5J 3M8, 403-422-0567.

LASER TECHNOLOGY IN BRITAIN

Opportunities for careers in laser technology in Britain are similar to those in the United States. For instance, lasers are used in manufacturing, surgery, communications, defense applications, and metrology (the science of measurement). "New uses are being found all the time," says Charles Childes, divisional secretary, Electronic Engineering Associates.

One British firm, Laser Scientific Services Ltd., in Cambridgeshire, England, has successfully developed the technology for laser cutting large aluminum aircraft panels in three dimensions. Technicians at the factory operate computer-programmed lasers to cut out the central slots and outer profile of a panel for a strategic aircraft.

Because of the laser's ability to produce a clean, sharp cut, the high quality cut edge of the panels needs no further machining and only limited deburring. For flat sheet work, tooling is not normally required. Complex geometric shapes can be cut in panels of various materials, using a sophisticated digitizer and computer-aided program system with direct numerical control. The thickness that can be cut depends on the material and the laser power.

Laser Scientific Services Ltd., which sells, installs, and maintains industrial laser systems, also uses a computer numerically controlled (CNC) laser to cut out lettering or logos with precise repeatability. Because the laser generates little heat, there is no distortion. A wide range of plastic, metal, laminate, and composite materials can be used. Other activities of the company include cutting, profiling, scribing, or drilling of ceramic components for the microelectronics industry. The company has set up a joint venture operation, using laser technology, in Denmark.

Addresses

For further information on lasers in the U.K., contact:

Electric Engineering Associates
  Leicester House
  8 Leicester St.
  London WC2H 7BN
  England
  (01 437 0678)

British Medical Association
  British Medical Association House
  Tavistock Square
  London WC 1
  England
  (01 387 4409)

Telecommunications Equipment
  Manufacturers Association
  Leicester House
  8 Leicester St.
  London WC2H 7BN
  England

LASER TECHNOLOGY IN AUSTRALIA

Australia has a strong background of research and development in mining, biomedical sciences, and scientific instrumentation—a background that is reflected in a high level of laser R&D, applications, and manufacture.

One key center for laser research and development has been established at Macquarie University in New South Wales. Here, under the direction of Professor Jim Piper, a laser physicist, researchers plan to investigate the basic science of laser physics while tailoring technology to specific applications and manufacturing within Australia. The center also plans to emphasize medical applications of lasers, both for therapeutic diagnostics and for surgery. Manufacturing products arising from the center are fabricated by Metalaser, a new company.

Australia currently is focusing its laser efforts in two fields: materials processing and medical lasers. Many of the industrial lasers are integrated with computer-numerically-controlled (CNC) systems, Jones says, so skills in computer programming are helpful. Medical lasers range from CO_2 lasers that are effectively miniaturized (a 9-cm laser puts out about 90 watts of power) to metal vapor lasers, developed in Australia. Copper and gold vapor lasers have been marketed; lasers using other metals are being tested. One joint project with Monash University and the University of Colorado in the United States involves a metal vapor laser that operates at room temperature, regarded as useful for biotechnology or biomedical processing instrumentation.

Nd:YAG lasers are being developed for opthalmology use by Laserex, an Australian company with some forty employees. Physicians at major hospitals and medical centers are using various types of lasers for dermatology, cancer surgery, gynecology, and photodynamic laser therapy.

Training Opportunities

Government officials continue to review the state of lasers in Australia. In April 1988 an important meeting was held to determine the strategy for Victoria, according to Dr. Alan J. Jones, assistant director, Exploitable Science and Technology, Department of Industry, Technology, and Commerce. "We have a shortage of skills," Jones says, "especially in the area of service and maintenance of existing laser systems. We're looking for qualified technical people."

A 1985 report, *Lasers Australia*, which Jones compiled, indicated Australia's need for more technical training. "University graduates have insufficient laser hands-on experience, and Ph.D. graduates would find no job satisfaction in intermittent repairs of equipment," Jones says. "There is an identifiable need for laser technologists who could gain their skills by one- to two-year associations with existing laser R&D facilities."

Both Jones and Dr. George L. Paul, director of the Centre for Industrial Laser Applications, School of Physics, the University of New South Wales, stress the need for more coursework in lasers. At the University of New South Wales, training in lasers and optics is offered in the third year of physics training. Other universities offer postgraduate courses related to lasers. A center for laser physics also operates at the Australian National University, Canberra, with a focus on mechanisms of laser fusion. Conventional Nd:YAG performance has been considerably improved during the course of this work, and this has found its way into the products of a new company, Electro-Optic Systems, which won a $30 million contract from the United States DOD for a series of its laser radar systems.

Research by postgraduate students is also being done at a number of other universities, according to Paul. However, he says, there aren't too many courses on modern optics being offered in Australia.

Some laser companies, Paul feels, believe it is better to retrain existing employees to use lasers than to hire specially trained technicians. "For instance, Laser Lab in Melbourne is successfully coupling U.S.-based lasers interfaced to up to 5-axis computer numerical controlled specialized cutting systems. The CNC software is designed by the Australian Numerical Control Association. Laser Lab, which supplies lasers for cutting, uses people they pulled from the factory floor and trained to use lasers," he explains.

"If young people want to work with lasers, then there may be job opportunities with similar companies. For instance, Metalaser, a new company that's in an expansion phase after purchasing Photon Sources, a California-based firm, is looking for more people."

Paul suggests that persons with training in optics who would like to get into the scientific instrument market are likely to find jobs. "There certainly are opportunities for trained people who come out to Australia," he explains.

It's a World Market

Paul emphasizes that from the Australian viewpoint, the laser market has become worldwide. "For those who want to get into new and exciting technologies, there are opportunities for vast profits," he says. "Teams of creative people are needed—people of diverse backgrounds, like chemists, physicists, metallurgists. Solving problems involves a wide range of skills, and that can be stimulating!"

Paul also points out that lasers, like some other technical fields in Australia, present niche market opportunities. "There are a number of these areas which Australian companies have identified and in which they're doing very well," he says. "For instance, Laser Lab in Melbourne is successful in using lasers for cutting. Laserex, another company, produces high-quality pocket-sized laser pointers. They supply about 50 percent of the world market, and they're doing extremely well. And Metalaser is doing a successful job in niche marketing for medical applications of metal vapor lasers."

Although there's been some shakeout in laser companies, Paul says, he feels the bigger companies are getting bigger. "Perhaps in the future, there will only be a small number of major laser companies," he says.

"But it's my belief that there are many areas of laser applications which have yet to be defined."

Additional Information

For more information on lasers and laser training in Australia, contact:

Centre for Lasers and Applications
c/o Professor Jim Piper
Macquarie University
North Ryde
New South Wales 2113
Australia
(02 805-7911)

Centre for Industrial
Laser Applications
Dr. George L. Paul
University of New South Wales
Kensington, NSW 2033
Australia
(02 697-4586)

Footscray Institute of Technology
Dr. Ken Peard
P.O. Box 64, Ballarat Rd.
Footscray, Victoria 3011
Australia
(03 688-4277)

Monash University
Dr. Rod Tobin
Clayton, Victoria 3168
Australia
(03 688-4277)

University of Queensland
Dr. David James
Director, Laser Applications
Laboratory
St. Lucia Queensland 4067
Australia
(07 377 2637)

Amtec Ltd. (an organization of the
Victorian government)
P.O. Box 322
Heidelberg, Victoria 3084
Australia
(03 480 0122)

Dr. Alan J. Jones
Assistant Director
Exploitable Science and
Technology
Department of Industry,
Technology, and Commerce
51 Allara Street
Canberra City 2600
Australia

ASSOCIATIONS

Because laser technology spans many areas, there are a number of trade associations involved with its varying aspects. Some of them have significant reduction in membership rates for qualified students. Most associations sponsor conferences or meetings that students can attend for reduced fees. Nearly all associations have journals, magazines, or other publications, and often these are available to students at discount rates. Some associations offer career guidance information. In addition, several associations offer videotapes that can be borrowed or rented.

There are many advantages to joining a professional association while you are still in school. Frequently, local or regional chapters have monthly meetings that student members can attend. These gatherings give you a chance to talk with established professionals, as well as those just beginning careers. The friendships you make will be extremely important.

Serving on a committee of such an association or society, even though you may still be a student, is well worth your time and effort. The informal contact with members, the chance to truly be a participant instead of just a spectator, the behind-scenes awareness of how a meeting or conference is actually put together—all these are good learning experiences. So is the chance to demonstrate leadership ability. While your motive for participation should be based on your genuine interest in the organization, nevertheless your activities will be watched by those already working in the field. They can help you with career questions and be a valuable source of information.

American Society for Laser Surgery and Medicine

This professional association of over 1,300 members in the United States and 20 additional countries is dedicated to exploring and applying laser technology to biology, medicine, and surgery. Members include physicists who develop devices, biomedical engineers who adapt them for practical purposes, safety officers who supervise workings of lasers, biologists who study the effects of laser energy on living tissue, and health professionals who treat patients with lasers.

Founded in 1980 the ASLMS has several classes of members. The Fellowship category includes scientists, physicians, physicists, veterinarians, podiatrists, dentists, and nurse-scientists. The Member category includes health care professionals, such as physical therapists, biomedical engineers, hospital administrators, and clinical nurses not eligible for Fellow status. Commercial Fellows are members involved in the commercial aspect of lasers as related to medicine, including development and marketing.

Listed by the American Medical Association as one of its recognized medical societies, ASLMS has formulated standards and guidelines for establishing safe, effective laser programs in hospitals and other institutions and has recommended standards for those who conduct both basic and advanced courses. ASLMS attempts to keep its membership informed about all postgraduate courses and training programs in laser biology, nursing, medicine, and surgery.

In association with other national, regional, international, and specialty laser societies, the ASLMS cooperates with the International Congress of Lasers in Surgery and Medicine, the Laser Institute of America, and the Society of Photo-Optical Instrumentation Engineers (SPIE) in organizing programs.

Lasers in Surgery and Medicine is the official journal of the American Society for Laser Medicine and Surgery. It carries proceedings of annual scientific meetings. A bimonthly newsletter includes news items, a calendar of events, courses and programs, and updates on government and insurance policies relating to laser biology, medicine, and surgery. The newsletter also includes a career opportunities section for job-seekers and employers.

For more information, write or call American Society for Laser

Medicine and Surgery, Inc., 813 Second Street, Suite 200, Wausau, WI 54401, (715) 845-9283.

Institute of Industrial Engineers

The Institute of Industrial Engineers, (IIE) is a major international professional society concerned with techniques, applications, and systems for industrial engineering. Members are concerned with the design, improvement, and installation of integrated systems of people, material, information, equipment, and energy—all the elements in the "productivity" equation.

IIE's 43,000 members belong to one or more of 23 separate divisions. Each division publishes newsletters that keep members up to date on new ideas and new developments in the field.

The institute has more than 200 senior chapters worldwide, plus over 100 university chapters at most universities offering industrial engineering curricula. All members receive *Industrial Engineering,* the monthly magazine. Books, periodicals, and microsoftware are also offered at membership discounts.

Two major conferences yearly (the Annual International Industrial Engineering Conference and Show and the Fall Industrial Engineering Conference), plus numerous seminars, continuing education programs, and workshops help members keep up with what's new professionally.

Through its career guidance program, which takes place primarily at the local level through established chapters, IIE acquaints thousands of young people each year with the opportunities and advantages of a career in industrial engineering. Through scholarships and fellowships, IIE rewards outstanding student members for scholastic excellence on both the graduate and the undergraduate levels.

For information on IIE, write or call Institute of Industrial Engineers, 25 Technology Park/Atlanta, Norcross, GA 30092, (404) 449-0460.

Laser Institute of America

Founded in 1968, the Laser Institute of America (LIA) is a nonprofit professional society for the advancement and promotion of laser technology and applications. It conducts continuing education courses, seminars, and technical symposia and offers a variety of educational materials and publications, including its official magazine, *Laser Topics.*

Within the institute, members have the opportunity to indicate their principal area of interest, choosing among medicine and biology; materials processing; inspection, measurement, and control; optical communications; information processing; holography; safety; scientific applications; imaging and display technology; and photochemistry/ spectroscopy.

Student chapters have been formed at a number of colleges, universities, and technical institutes. These chapters work closely with professionals. For instance, LIA's student chapter at Pasadena City College sponsored *Laser Expo '87,* a comprehensive format of seminars and displays highlighting laser electro-optics technology. The Expo, funded through grants and private efforts, attracted more than 2,000 attendees.

LIA sponsors one of the major annual conferences on lasers— ICALEO, the International Congress on Applications of Lasers and Electro-Optics—in cooperation with the American Society for Laser Medicine and Surgery; the Society of Manufacturing Engineers; the International Society of Podiatric Laser Surgery; the American Society of Metals, International; the Midwest Bio-Laser Institute; the High Temperature Society of Japan; the Western Institute for Laser Treatment; the Japan Laser Processing Society; the Japan Society for Laser Technology; and IFS Conferences, Ltd.

ICALEO features four simultaneous technical conferences: Laser Materials Processing, Laser Research in Medicine, Optical Methods in Flow and Particle Diagnostics, and Electro-Optic Sensing and Measurement.

Throughout the year LIA works closely with other laser-related organizations to coordinate various short courses for professionals. For instance, in 1988 its laser electro-optics offerings included "Fundamentals and Applications of Lasers"; "Modern Experimental Spectroscopy"; "Laser Safety"; "Hazards, Inspection and Control"; "Advanced Industrial Laser Safety Officer Training"; and "Medical

Laser Safety Officers Training". These three- or four-day courses were offered at various locations around the country.

For information on LIA, including the location of student chapters, write or call Laser Institute of America, 5151 Monroe Street, Toledo, OH 43623, (419) 882-8706.

Optical Society of America

Founded in 1916, the Optical Society of America has nearly 10,000 individual members, including scientists, engineers, and technicians from the United States and 50 other countries. Members work in industry, educational institutions, and government agencies, and include a number of Nobel laureates.

More than 60 companies with an interest in optics have pledged corporate support to the mission statement of the OSA: "To increase and diffuse the knowledge of optics, to promote the common interests of investigators of optical problems, of designers and users of optical apparatus of all kinds, and to encourage cooperation among them."

Among OSA publications are its five peer-reviewed journals and its news magazine, *Optics News.* Each is devoted to a specific aspect of optical science or technology. They are *Applied Optics, Journal of Lightwave Technology, Journal of the Optical Society of America A: Optics and Image Science, Journal of the Optical Society of America B: Optical Physics,* and *Optics Letters.*

In addition, OSA publishes three translated journals in English: *Chinese Physics—Lasers, Optics and Spectroscopy,* and *Soviet Journal of Optical Technology.*

Information on lasers and laser physics is generally found in *JOSA B* and in *Applied Optics.* Information on topics such as fiber and cable technologies is covered in *Journal of Lightwave Technology,* published jointly by OSA and the Institute of Electrical and Electronics Engineers. This publication presents advances in the science, technology, and engineering of optical guided waves. *Optics News* includes ongoing coverage of developments in instrumentation and systems applications for lasers and optical fibers. *Optics Letters* includes new results in optics research, including fiber-optics technology.

More than 20 digests are published annually by OSA. These cover all OSA-sponsored and cosponsored conferences and meetings.

Although OSA does not sponsor student chapters at universities, individual students can join the society at a reduced rate of $18 per year. Student membership entitles persons to discounts on all publications and journals. In addition, students who are members can attend all OSA-sponsored conferences at reduced rates. If they wish, they can be put in touch with local OSA chapters and can attend local meetings.

For additional information on the Optical Society of America, write or call its executive office, 1816 Jefferson Place NW, Washington, DC 20036, (202) 223-8130.

Society of Manufacturing Engineers

The Society of Manufacturing Engineers (SME) is one of the largest professional engineering associations in the world, with more than 80,000 members in 69 countries. It has more than 10,000 student members and more than 125 student chapters.

Within SME, various associations are concerned with different aspects of manufacturing: Robotics International (RI/SME), Computer & Automated Systems Association (CASA/SME), Association for Finishing Processes (AFP/SME), North American Manufacturing Research Institute (NAMRI/SME), and Machine Vision Association (MVA/SME).

As the umbrella organization for these associations, SME annually sponsors more than 150 special programs, 50 technical conferences, 40 expositions, and 200 symposia and workshops. It publishes technical papers on various subjects that can be purchased either individually or (often) in various collections. An on-line electronic database search, available through the society's library service, allows members and nonmembers to access papers on desired topics quickly.

SME has a Laser Council, composed of individuals experienced in the use of lasers in manufacturing. Meeting several times throughout the year, the Laser Council plans SPOT, an intensive annual conference that features papers describing laser technology and applications.

In addition, SME has produced a 30-minute videotape on lasers as they're used in today's manufacturing plants. The tape features interviews with industrial leaders from small, medium, and large companies who explain the use of lasers to meet their firms' specific manufacturing needs. It's part of a series of *Manufacturing Insights.*

For more information on SME, write or call the Society of Manufacturing Engineers, One SME Drive, Dearborn, MI 48121, (313) 271-1500. Membership rates include a subscription to *Manufacturing Engineering,* SME's monthly magazine.

Society of Photo-Optical Instrumentation Engineers

SPIE (Society of Photo-Optical Instrumentation Engineers) is a nonprofit society dedicated to advancing engineering and scientific applications of optical, electro-optical, and optoelectronic instrumentation, systems, and technology. Its members are scientists, engineers, and users interested in these technologies. SPIE uses publications and conferences to communicate new developments and applications to the scientific, engineering, and user communities.

Annual membership dues are $60. Members receive SPIE's monthly journal, *Optical Engineering,* which presents technical papers reporting on technologies relating to the engineering, design, production, and applications of optical, electro-optical, optoelectronic, fiber-optic, laser, and photographic components and systems.

Student membership is available at a reduced rate, and two student chapters currently exist: at California State Polytechnic University in San Luis Obispo and at the University of Houston/Clear Lake. Several other petitions for student chapters are pending.

SPIE sponsors a number of conferences on many topics. Typical of conferences related to lasers is SPIE's O-E/LASE '88, a week-long conference covering opto-electronics and laser applications in science and engineering. Separate symposia were held on Lasers and Optics; Innovative Science & Technology, with related conferences on Optical Signal Processing; Laser Spectroscopy—Techniques, Applications, Data Bases, and Equipment; Medical Applications of Lasers, Fiber Optics, and Electro-optics; and Electronic Imaging and Optical Mass

Data Storage. In all, 43 technical conferences and 104 education courses were presented.

More than 150 exhibitors demonstrated products, instruments, and services used in optics, lasers, imaging, spectroscopy, interferometry, optical mass data storage, optical signal processing, and medicine.

Like all the society's conferences, O-E/LASE '88 was open—that is, attendance was not restricted to members only or to professionals working in the field. Complete proceedings from the conferences were published and available for purchase from SPIE.

For further information on joining SPIE or on its conferences and publications, write SPIE—The International Society for Optical Engineering, PO Box 10, Bellingham, WA 98227-0010, (206) 676-3290.

RECOMMENDED READING AND RESOURCES

Because laser technology is a rapidly changing field, you'll want to keep up with news and developments. You can find basic information on lasers in reference books and in books on library shelves. However, since new information is constantly being discovered about lasers and their uses, you will want to have the most current publications.

Periodicals

One way to get up-to-date information is to use library indexes. You've probably already used *Reader's Guide to Periodical Literature,* an index commonly found in public libraries that tracks articles in many popular magazines by subject matter. Although *Reader's Guide to Periodical Literature* is a good place to start, the material referred to may not be technical enough for your needs.

Fortunately, most libraries have companion indexes, organized along similar lines. *Business Periodicals Index* is a similar reference work that focuses on magazines and publications of interest in financial and economic fields. You can look up "Lasers" in recent volumes of BPI and can then go to the individual publications for particular articles. Stories and news you would find about lasers might include material on company profits, mergers and acquisitions, lawsuits, or future sales.

One way in which you can keep up with technical advances in laser technology is to use *Applied Science and Technology Index,* another specialized index available at most libraries. Here, the broad topic of

"Lasers" is subdivided into such headings as "Lasers—Industrial Applications," "Laser Printers," or "Lasers—Measurement Methods."

At the beginning of a bound volume of each of these indexes you will find a list of the periodicals covered by that particular index along with an address and subscription rate for each. You will find that certain publications show up frequently in these indexes. If your library does not subscribe to a magazine you wish, you may want to write the publisher directly, enclosing money and a large, self-addressed, stamped envelope for a recent copy. Looking closely at the magazine can help you decide whether you may want to become a subscriber or whether you want to ask your library for help in locating more copies.

In large metropolitan areas such as Chicago, many corporations and research laboratories maintain their own libraries on topics of interest in their particular field. You will find special books and magazines on food and nutrition, for example, at the Quaker Oats Research Laboratories in suburban Barrington, Illinois, while Kemper Insurance Company in suburban Long Grove, Illinois, has a collection that includes *Best Insurance Reports,* the *West Reporter System* (law digests), a collection on alcoholism, and various proceedings and transactions of societies and conferences. Kemper's subject strengths are insurance and insurance law, law in general, and management.

If you are located in an area that has companies that manufacture, sell, or service equipment in electronics, physics, or telecommunications, those businesses, especially if they are large, may well have corporate libraries. Your own public library reference librarian will know what's available in your area, whether or not you can use those corporate libraries for study and reading, and whom to call at a particular location to find out what periodicals and books they have. Often, if a magazine is expensive or so specialized that a public library doesn't subscribe to it, these corporations will have copies you can look at.

Many libraries in metropolitan areas also have a service under which they share information. In the Chicago area, a number of public libraries belong to Central Serials Service, an interlibrary consortium which makes photocopies of articles you can't find at your local library and sends them to your local librarian—at no charge to you. There are a number of rules you must obey to use this service; most of them have to do with the frequency with which you want articles from a particular publication. Copyright laws permit limited reproduction of copyrighted material, so librarians want to be sure you do not ask for more than you're entitled to.

Still another way in which you may get material on lasers is through computer searching. Reference librarians often are tied in with various databases. Because you are a card-holder at your local library, the library may be willing to do a certain number of free or low-cost searches per year on topics related to lasers. If you are requesting an on-line search, check first with your librarian to see whether you will be charged. Generally, you can probably find more than enough material in your field of interest by reading copies of the publications listed below.

Advanced Materials & Processes Incorporating Metal Progress
American Society for Metals
Metals Park, OH 44073

Aerospace America
1633 Broadway
New York, NY 10019

AIAA Journal
American Institute of Aeronautics and Astronautics, Inc.
1633 Broadway
New York, NY 10019

American Machinist & Automated Manufacturing
PO Box 502
Hightstown, NJ 08520

Analytical Chemistry
American Chemical Society
PO Box 3337
Columbus, OH 43210

Applied Optics
American Institute of Physics
335 East 45th Street
New York, NY 10017

ASTM Standardization News
American Society for Testing and Materials
1916 Race Street
Philadelphia, PA 19103

Automotive Engineering
Society of Automotive Engineers
400 Commonwealth Drive
Warrendale, PA 15096

Aviation Week & Space Technology
PO Box 1505
Neptune, NJ 07753

Byte
PO Box 328
Hancock, NH 03449

Communication News
Edgell Communications
124 South First Street
Geneva, IL 60134

Computer
IEEE Computer Society
10662 Los Vaqueros Circle
Los Alamitos, CA 90720

Design News
Cahners Publishing Company
270 Saint Paul Street
Denver, CO 80206

Electronic Engineering
Morgan Grampian, Ltd.
309 Calderwood Street
Woolwich, London SE18 6QH
England

Electronics
CN 808
Martinsville, NJ 08836

The Engineer
  Morgan Grampian, Ltd.
  Royal Sovereign House
  40 Beresford Street
  London SE18 6BQ
  England

Fiberoptic Product News
  Gordon Publications, Inc.
  13 Emery Avenue
  PO Box 1952
  Dover, NJ 07801-0952

Fortune
  Time, Inc.
  Time & Life Building
  Rockefeller Center
  New York, NY 10020-1393

High Technology
  Infotechnology Publishing Corporation
  38 Commercial Wharf
  Boston, MA 02110

Holography News
  PO Box 9796
  Washington, DC 20016

IEEE Computer Graphics and Applications
  IEEE Service Center
  10662 Los Vaqueros Circle
  Los Alamitos, CA 90720-2578

IEEE Journal of Quantum Electronics
  IEEE Service Center
  445 Hoes Lane
  Piscataway, NJ 08854-4150

IEEE Journal of Solid-State Circuits
  IEEE Service Center
  445 Hoes Lane
  Piscataway, NJ 08854-4150

IEEE Transactions on Industry Applications
  IEEE Service Center
  445 Hoes Lane
  Piscataway, NJ 08854-4150

Industrial Engineering
Institute of Industrial Engineers
25 Technology Park/Atlanta
Norcross, GA 30092

Industrial & Engineering Chemistry Research
American Chemical Society
PO Box 3337
Columbus, OH 43210

The Industrial Robot
IFS Publications, Ltd.
35-39 High Street
Kempston, Bedford MK42 7BT
England

ISA Transactions
Instrument Society of America
67 Alexander Drive
Research Triangle Park, NC 27709

Journal of the American Ceramic Society
65 Ceramic Drive
Columbus, OH 43214

Journal of Applied Physics
American Institute of Physics
335 East 45th Street
New York, NY 10017

Journal of Biomechanical Engineering
American Society of Mechanical Engineers
345 East 47th Street
New York, NY 10017

Journal of the Electromechanical Society
10 South Main Street
Pennington, NJ 08534-2896

Journal of Engineering for Gas Turbines and Power
American Society of Mechanical Engineers
345 East 47th Street
New York, NY 10017

Journal of Metals
420 Commonwealth Drive
Warrendale, PA 15086

Journal of the Optical Society of America A: Optics and Image Science
American Institute of Physics
500 Sunnyside Boulevard
Woodbury, NY 11797

Journal of the Optical Society of America B: Optical Physics
American Institute of Physics
500 Sunnyside Boulevard
Woodbury, NY 11797

Journal of Physical Chemistry
American Chemical Society
1155 16th Street NW
Washington, DC 20036

Journal of Physics D: Applied Physics
Physics Trust Publications
823-825 Bath Road
Bristol BS4 5NU
England

Journal of Physics E: Scientific Instruments
Physics Trust Publications
823-825 Bath Road
Bristol BS4 5NU
England

Journal of Solar Energy Engineering
American Society of Mechanical Engineers
345 East 47th Street
New York, NY 10017

Laser Focus
PO Box 1111
Littleton, MA 01460

Lasers & Optronics
23868 Hawthorne Boulevard
Torrance, CA 90505-5908

Light Metal Age
Roy Fellom, Jr.
693 Mission Street
San Francisco, CA 94105

Machine Design
  Penton/IPC
  Penton Plaza
  Cleveland, OH 44114

Manufacturing Engineering
  Society of Manufacturing Engineers
  PO Box 930
  Dearborn, MI 48121

Material Handling Engineering
  PO Box 95759
  Cleveland, OH 44101

Metal Finishing
  1 University Plaza
  Hackensack, NJ 07601

Metallurgia
  Fuel and Metallurgical Journals, Ltd.
  Queensway House
  2 Queensway
  Redhill, Surrey RH1 1QS
  England

Modern Casting
  American Foundrymen's Society, Inc.
  Golf and Wolf Roads
  Des Plaines, IL 60016-4090

Modern Machine Shop
  Gardner Publications, Inc.
  6600 Clough Pike
  Cincinnati, OH 45244-4090

Modern Materials Handling
  Cahners Publishing Company
  275 Washington Street
  Newton, MA 02158

Modern Metals
  211 East Chicago Avenue
  Chicago, IL 60611

New Scientist
New Science Publications
Commonwealth House
1-19 New Oxford Street
London WC1A 1NG
England

Nuclear Engineering International
Business Press International, Ltd.
Oakfield House
Perrymount, Haywards Heath, Sussex RH16 3BR
England

OE Reports
The International Society for Optical Engineering
PO Box 10
Bellingham, WA 98227-0010

Optical Engineering
Society of Photo-Optical Instrumentation Engineers
PO Box 10
Bellingham, WA 98227-0010

Optics News
Optical Society of America
1816 Jefferson Place NW
Washington, DC 20036

Packaging
Cahners Publishing Company
270 Saint Paul Street
Denver, CO 80206-5191

Physics Today
AIP S/F Division
500 Sunnyside Boulevard
Woodbury, NY 11797

Plasma Physics and Controlled Fusion
Pergamon Press, Inc.
Maxwell House, Fairview Park
Elmsford, NY 10523

Polymer Engineering and Science
Society of Plastics Engineers Inc.
14 Fairfield Drive
Brookfield Center, CT 06805

Radio-Electronics
  PO Box 5515
  Boulder, CO 80321-5115

RCA Review
  RCA Laboratories
  Princeton, NJ 08540

Review of Scientific Instruments
  American Institute of Physics
  335 East 45th Street
  New York, NY 10017

Robotics Engineering
  174 Concord Street
  Peterborough, NH 03458

Rock Products
  300 West Adams Street
  Chicago, IL 60606

Scientific American
  PO Box 5969
  New York, NY 10017

Science
  American Association for the Advancement of Science
  1333 H Street NW
  Washington, DC 20005

Solid State Technology
  875 Third Avenue
  New York, NY 10022

Technology Review
  Room 10-140, MIT
  Cambridge, MA 02139

Telecommunications
  685 Canton Street
  Norwood, MA 02062

Tooling & Production
  International Thomson Industrial Press, Inc.
  6521 Davis Industrial Parkway
  Solon, OH 44139

Vacuum
  Pergamon Press, Inc.
  Maxwell House, Fairview Park
  Elmsford, NY 10523

Books

In order to increase your understanding of how lasers work and of laser applications in various fields, it will be helpful for you to do as much reading as possible. If books suggested below are not available through your school or public library, a librarian can probably arrange to borrow them through interlibrary loan.

The reading level and complexity of ideas explored in the books on this list vary considerably—that is, you cannot necessarily tell from a book's title just how difficult it will be. One good way to be sure you understand basic principles of lasers is to begin reading at a level slightly below your present knowledge of physics, optics, or electronics. The books in the following list that are marked with an asterisk may be a good place to start, since they are somewhat easier.

Because knowledge about lasers and laser applications is growing rapidly, as you progress in your self-study beyond understanding the basic technology, you will want to pay attention to publication dates, so that the information you obtain is as current as possible. That is why keeping up with trade journals and magazine articles is also essential.

The following list is, of course, by no means complete. Your reference librarian can help you find other books about lasers that you may enjoy. You may want to consult your math or science teacher or even your school district's curriculum consultant in science. Another way you can find more books is to write to one of the colleges or universities listed in chapter 9. Address your letter to the dean of the engineering school, the head of the physics department, or to someone holding a comparable position. Tell them you would like to know more about lasers and ask them to recommend a reading list for high school students.

By the time you've sampled 8 or 10 books from the list below, you should know how serious you are about working with lasers. Like many people, you will find the story of laser light and its applications fascinating and challenging!

*Barrett, N.S., *Lasers and Holograms.* London, England: Franklin Watts, 1985.

*Bender, Lionel. *Lasers in Action.* New York, New York: The Bookwright Press, 1985.

Benedict, Gary F. *Nontraditional Manufacturing Processes.* New York, New York: Marcel Dekker, Inc., 1987. (contains excellent chapter on lasers in manufacturing)

Burkig, Valerie. *Photonics, the New Science of Light.* Hillside, New Jersey: Enslow Publishers, Inc., 1986.

*Burroughs, William. *Lasers.* New York, New York: Warwick Press, 1982.

Hallmark, Clayton L. *Lasers, the Light Fantastic.* Blue Ridge Summit, Pennsylvania: Tab Books, Inc., 1987.

*Johnson, James. *Lasers.* Milwaukee, Wisconsin: Raintree Publishers, 1981.

Kettlekamp, Larry. *Lasers, the Miracle Light.* New York, New York: William Morrow and Company, 1979.

Masten, Larry B. and Billy R. *Understanding Optronics.* Dallas, Texas: Texas Instruments Learning Center, 1981.

McAleese, Frank G. *The Laser Experimenter's Handbook.* Blue Ridge Summit, Pennsylvania: Tab Books, Inc., 1979.

*McKie, Robin. *Lasers.* New York, New York: Franklin Watts, 1983.

Mims, Forrest M., III. *Lasers (The Incredible Light Machines).* New York, New York: David McKay Company, Inc., 1977.

Muncheryan, Hrand M. *Principles & Practices of Laser Technology.* Blue Ridge Summit, Pennsylvania: Tab Books Inc., 1983.

*Olesky, Walter. *Lasers.* Chicago, Illinois: Children's Press, 1986.

Safford, Edward L., Jr. *The Fiberoptics & Laser Handbook.* Blue Ridge Summit, Pennsylvania: Tab Books, Inc., 1984.

*Schneider, Herman. *Laser Light*. New York, New York: McGraw-Hill Book Company, 1978.

Seippel, Robert G. *Fiber Optics*. Reston, Virginia: Reston Publishing Company, Inc., 1984.

Siegman, Anthony E. *Lasers*. Mill Valley, California: University Science Books, 1986.

*Stevens, Lawrence. *Laser Basics—An Introduction for Young People*. Englewood Cliffs, New Jersey: Prentice-Hall, Inc., 1985.

Wenyon, Michael. *Understanding Holography*. New York, New York: Arco Publishing, Inc., 1985.

Videos

A videotape explaining laser holography, suitable for junior high and high school classes, is available for purchase ($39.95) or preview from the Thomas Alva Edison Foundation, 2100 West Ten Mile Road, Southfield, Michigan 48075. Additional resources from this Foundation include a booklet, "Laser Holography, Experiments You Can Do." $3.00 plus postage/handling.

Two videos about lasers in manufacturing are available from the Society of Manufacturing Engineers, One SME Drive, P.O. Box 930, Dearborn, Michigan 48121, 313-271-1500. *Lasers in Manufacturing—A New Look,* a tape which complements an older tape, *Lasers in Manufacturing,* illustrates how lasers are used to machine, heat treat, and perform alignment and calibration functions on a variety of materials and equipment. Included in the 30-minute tape is footage of in-plant scenes at five manufacturing firms.

The Society also offers a video describing careers in manufacturing engineering. Details are available from the Reference Publications Department, x 416.